The Client Role in Successful Construction Projects

The Client Role in Successful Construction Projects is a practical guide for clients on how to initiate, procure and manage construction projects and developments. This book is written from the perspective of the client initiating a construction project as part of a business venture and differs from most available construction literature which can externalise the client as a risk to be managed by the design team. The book provides a practical framework for new and novice clients undertaking construction, giving them a voice and enabling them to:

- Understand the challenges that they and the project are likely to face.
- Communicate and interact effectively with key stakeholders and professionals within the industry.
- Understand in straightforward terms where they can have a positive impact on the project.
- Put in place a client-side due diligence process.
- Reduce their institutional risk and the risk of project failure.
- Discover how their standard models are able to co-exist and even transfer to a common client-side procedure for managing a construction project.

Written by clients, for clients, this book is highly recommended not only for clients, but for construction industry professionals who want to develop their own skills and enhance their working relationship with their clients. A supporting website for the book will be available, which will give practical examples of the points illustrated in the book and practical advice from specialists in the field.

Jason Challender has acquired nearly 30 years' client-side experience in the UK construction industry and procured numerous successful major construction programmes during this time. He is Director of Estates and Facilities at the University of Salford, a member of its Senior Leadership Team and is responsible for overseeing a large department of approximately 350 estates and construction related staff. He is also a construction researcher with ten published academic journal and conference papers in recent years, all of which have been dedicated to his studies around trust and collaborative procurement in the construction industry. Furthermore, he has previously participated as a book reviewer for Wiley. He has also attended many national construction and institutional conferences as a guest speaker over the years and is a Fellow and Board Director of the Royal Institution of Chartered Surveyors.

Russell Whitaker is a Chartered Surveyor, an engineer and construction purchasing specialist and runs his own practice in Guildford. Russell was a senior executive and former Vice Principal in many public and private organisations with over 30 years' managing complex multi-faceted development and construction projects and facilities staff before starting his own property consultancy. He has advised many prestigious clients including Southbank Arts Centre, Royal Botanic Gardens Kew and the Royal Automobile Club on project initiation, pre-development and procurement. Russell has strong academic links as a Visiting Lecturer and Course Leader for over 20 years in Building Project Management and Procurement and Professional Project Management at the Sir John Cass Business School, City University. He is a conference speaker specialising in the client role in managing 'front-end' project risk.

The Client Role in Successful Construction Projects

Jason Challender and Russell Whitaker

LONDON AND NEW YORK

First published 2019
by Routledge
2 Park Square, Milton Park, Abingdon, Oxon OX14 4RN

and by Routledge
52 Vanderbilt Avenue, New York, NY 10017

Routledge is an imprint of the Taylor & Francis Group, an informa business

© 2019 Jason Challender and Russell Whitaker

The right of Jason Challender and Russell Whitaker to be identified as authors of this work has been asserted by them in accordance with sections 77 and 78 of the Copyright, Designs and Patents Act 1988.

All rights reserved. No part of this book may be reprinted or reproduced or utilised in any form or by any electronic, mechanical, or other means, now known or hereafter invented, including photocopying and recording, or in any information storage or retrieval system, without permission in writing from the publishers.

Trademark notice: Product or corporate names may be trademarks or registered trademarks, and are used only for identification and explanation without intent to infringe.

British Library Cataloguing-in-Publication Data
A catalogue record for this book is available from the British Library

Library of Congress Cataloging-in-Publication Data
Names: Challender, Jason, author. | Whitaker, Russell, author.
Title: The client role in successful construction projects /
Jason Challender, Russell Whitaker.
Description: Abingdon, Oxon ; New York, NY : Routledge is an imprint of the Taylor & Francis Group, an Informa Business, 2019. |
Includes bibliographical references and index.
Identifiers: LCCN 2018058264 | ISBN 9781138058200 (hbk) |
ISBN 9781138058217 (pbk) | ISBN 9781315164458 (ebk)
Subjects: LCSH: Building–Superintendence. | Construction industry–Customer services. | Customer relations–Management. | Contractors–Selection and appointment. | Consumer satisfaction. | Leadership.
Classification: LCC TH438 .C447 2019 | DDC 690.068–dc23
LC record available at https://lccn.loc.gov/2018058264

ISBN: 978-1-138-05820-0 (hbk)
ISBN: 978-1-138-05821-7 (pbk)
ISBN: 978-1-315-16445-8 (ebk)

Typeset in Goudy
by Newgen Publishing UK

Contents

List of figures	xiii
List of tables	xv
Foreword by Miles Wade	xvi
Foreword by Peter McDermott	xviii
Preface	xxi
Acknowledgements	xxiv

PART 1
An introduction into the construction industry and construction clients   1
JASON CHALLENDER

1 **Introduction**   3

2 **A model of construction clients and their projects**   8
   2.1 *Clients and the process of construction* 8
   2.2 *The 'client voice'* 9
   2.3 *The complexity of clients* 10
   2.4 *Client requirements through project briefs* 12
   2.5 *The transformational nature of client projects* 15
   2.6 *Project and client requirements* 17
   2.7 *Summary* 18

3 **The unique role of clients in the context of the construction industry**   21
   3.1 *The bespoke nature of the construction industry* 21
   3.2 *The construction industry and recent reforms* 23
   3.3 *The involvement of clients in the procurement of design and construction services* 23
   3.4 *Overall context of collaborative working and partnering within the construction industry* 26

  3.5 Alternative procurement methods for clients on construction projects 28
  3.6 Summary 31

4 The importance of leadership for construction clients 35
  4.1 Introduction 35
  4.2 Leadership identity and relevance for construction clients 35
  4.3 Leadership challenges for construction clients 37
  4.4 Key components of leadership for construction clients 37
  4.5 Leadership styles as applied to construction clients 38
  4.6 The importance of communication for construction clients in leadership 39
  4.7 Adaptability of leadership styles for construction clients 41
  4.8 The relationship between poor client leadership and project failure 41
  4.9 Summary 42

5 Governance considerations for construction clients 45
  5.1 Introduction 45
  5.2 Project controls 45
  5.3 The importance of project/programme boards 45
  5.4 Gateway processes for project approval and business cases 46
  5.5 Summary 49

6 Selection and appointment processes for construction clients 51
  6.1 Introduction 51
  6.2 The importance of the contractor selection process 52
  6.3 Articulation of the problem for selecting contracting partners from the perspective of construction clients 53
  6.4 A historical perspective of the problem 54
  6.5 Risk 55
  6.6 Benchmarking 56
  6.7 Pre-qualification models and methodologies 62
  6.8 Designing a new way for construction clients to select their contracting partners 64
  6.9 The quality of pre-qualification processes and their influence on project success 66
  6.10 Summary and conclusions 68

7 The 'Intelligent Client': a model of procurement built on relationship management between construction clients and the supply chain 74
  7.1 Introduction 74
  7.2 What do we mean by collaborative strategies? 74

Contents ix

    7.3 *An awareness for construction clients on issues around collaborative procurement strategies and trust 75*
    7.4 *Trust as a collaborative necessity for construction clients 76*
    7.5 *Potential benefits of trust for construction clients; incentives to trust 79*
    7.6 *Research findings and discussion 80*
    7.7 *Implications for construction clients in managing requirements and expectations for collaboration and trust 82*
    7.8 *Conclusions and recommendations 84*

8 Critical success factors for construction clients   89
    8.1 *Introduction 89*
    8.2 *The articulation of the problem for construction clients 91*
    8.3 *Understanding what skills construction clients require 92*
    8.4 *Success criteria on projects for construction clients 93*
    8.5 *Research study to identify the success factors for construction clients 98*
    8.6 *Analysis and reflection of the study on critical success factors for construction clients 99*
    8.7 *The key human skills that influence the performance of construction clients 102*
    8.8 *The importance of individual qualities of construction clients for project success 103*
    8.9 *Summary 104*

9 The relevance of professional ethics for construction clients   107
    9.1 *Introduction 107*
    9.2 *What are professional ethics? 107*
    9.3 *The importance of professional ethics for construction clients 108*
    9.4 *Codes of ethics for construction clients 110*
    9.5 *How should construction clients recognise unethical practices? 111*
    9.6 *The need for construction clients to uphold cultural values when procuring projects 112*
    9.7 *Governance and regulation of professional ethics 113*
    9.8 *Environmental ethics 115*
    9.9 *Summary and conclusions 115*

10 The influence of construction clients on motivating project teams   117
    10.1 *Why should construction clients be concerned about project team motivation? 117*
    10.2 *The bespoke and unique nature of the construction industry 118*
    10.3 *Factors which affect motivation levels of project teams 119*
    10.4 *Communication as a motivational factor 120*
    10.5 *The benefits of training and education 122*

10.6 Cultural factors affecting levels of motivation for construction related staff *124*
10.7 The use of financial incentives as a motivational management tool *126*
10.8 Summary and conclusions *127*

11 Developing a construction client toolkit, as a practical guide for managing projects — 132

11.1 Introduction *132*
11.2 Planning and devising the toolkit *133*
11.3 Feedback and evaluation of the toolkit from the perspectives of construction clients *133*
11.4 Ensuring and monitoring performance throughout the life of projects: general project directory and checklist *134*
11.5 The documentation that construction clients need to consider at pre-construction stages *134*
11.6 Managing documentation and construction processes following the appointment of contractors *139*
11.7 The documentation and processes that construction clients need to consider in the construction phases of projects *139*
11.8 The documentation and processes that construction clients need to consider in the post-construction phases of projects *140*
11.9 Conclusion *148*

12 Reflections, overview and summary of key points of Part 1 — 149

12.1 Overall summary and recommendations *149*

Appendix 1A Project proposal (Gateway 1) — 153

Appendix 1B Business case (Gateway 3) template — 155

Appendix 1C Example of a partnering charter — 168

Appendix 1D Example of a contractor competency questionnaire — 169

Appendix 1E Example of a health and safety contractor's handbook — 175

Appendix 1F Example of a project execution plan (PEP) — 184

PART 2
A construction risk management model for clients — 205
RUSSELL WHITAKER

13 Executive summary briefing — 207

14 Construction clients, business propositions and corporate construction risk 210
- 14.1 The construction client 210
- 14.2 Construction clients and value propositions 211
- 14.3 Client and construction team behaviours 212
- 14.4 Risk management principles for clients in construction 215
- 14.5 Summary: the need for a professional construction client 219

15 Unique client contributions to construction 221
- 15.1 Preamble 221
- 15.2 The professional construction client 221
- 15.3 The four unique contributions a client can make to a construction client 222
- 15.4 Project success and client satisfaction 229
- 15.5 Summary: the client as a unique contributor to project success 229

16 Reducing corporate risk using the construction risk management model 231
- 16.1 Risk, construction and clients 231
- 16.2 Client contributions to risk reduction 232
- 16.3 Navigating around the client-side Construction Enterprise Risk Management (CERM) model 233
- 16.4 Summary: managing risk using our client-side risk management model 235

16a Business Concept Development Stage 236
- 16a.1 Purpose 236
- 16a.2 Organisations and planning 236
- 16a.3 Planned change and managed impact 237
- 16a.4 Why construction change is different 237
- 16a.5 Business Concept Development Stage outputs 238
- 16a.6 Job book: Business Concept Development Stage 240
- 16a.7 Summary: Business Concept Development Stage 246

16b Corporate Client/Delivery Capability/Transformation Stage 247
- 16b.1 Purpose 247
- 16b.2 Corporate Client/Delivery Capability/Transformation Stage 247
- 16b.3 Job book: Corporate Client/Delivery Capability/Transformation Stage 248
- 16b.4 Summary: Corporate Client/Delivery Capability/Transformation Stage 258

16c Outcomes Delivery Stage 259
- 16c.1 Purpose 259

    16c.2 *Outcomes Delivery Stage* 259
    16c.3 *Job book: Outcomes Delivery Stage* 261
    16c.4 *Preparing for design stage, benchmarking and design liability* 265
    16c.5 *Appointment of contractors and start on site* 270
    16c.6 *Summary: Outcomes Delivery Stage* 271

16d Three Year In Stage   272
    16d.1 *Purpose* 272
    16d.2 *Three Year In Stage* 272
    16d.3 *Job book: Three Year In Stage* 274
    16d.4 *Summary: Three Year In Stage* 279

17 Key points: client risk management and the risk management model   281

Appendix 2A Client-side CERM model   285

Appendix 2B RIBA Outline Plan Stage of Work   286

    Index   287

Figures

2.1	Integral components of the client body	9
2.2	The influence of the clients' organisation and experience on information processing at briefing stages	11
2.3	Basic structure of a construction client model	15
2.4	The client's business operation model with influence mapping	18
2.5	Internal and external influences on a client organisation	19
3.1	The iron triangle of cost, time, quality and scope	22
3.2	Diagram illustrating traditional procurement structure	25
3.3	Traditional procurement route under RIBA Plan of Work	25
3.4	Design and build procurement structure	27
3.5	Design and build procurement route under RIBA Plan of Work	27
3.6	Benefits for clients in early integration of contractors	29
5.1	Chart illustrating an example of a client approval process across a hierarchy of different project/programme boards	47
5.2	Flowchart to illustrate a gateway approval process	50
6.1	Performance criteria and subcriteria	60
7.1	Conceptual framework between trust and success	77
7.2	Flowchart to illustrate the influence of trust on improved project performance	78
7.3	Scatter diagram illustrating the influence of trust building mechanisms (independent variable) to the level of trust generated on projects (dependent variable)	80
8.1	Management skills necessary at various levels of an organisation	93
11.1	General project checklist	135
11.2	Example of a project dashboard template	143
14.1	Project lifecycle BS6079	213
14.2	Risk dynamics in construction projects	216
14.3	Risk status matrix and the use of ERM (CERM)	218
15.1	Client behaviours and the broad outputs of an ERM system for construction	223
15.2	Client-side Construction ERM (CERM) model	227
15.3	Relationship between CERM model against RIBA Outline Plan Stage and BS6079	228

16.1	Client-side CERM model	234
16a.1	Business Concept Development Stage	236
16a.2	Change and business boundaries	238
16a.3	Business Concept Development Stage	244
16b.1	Corporate Client/Delivery Capability/Transformation Stage	247
16b.2	Outer and inner circle financial controls	254
16c.1	Outcomes Delivery Stage	259
16d.1	Three Year In Stage	272

Tables

2.1	Client brief checklist	14
5.1	Example of financial levels of authority for a client organisation	48
7.1	Collaboration outputs, outcomes, key activities and expectations of clients and their construction partners	83
8.1	Content analysis and qualitative themes	94
8.2	Percentage rate of theme titles in relation to the Skill Approach	99
9.1	Top 11 most frequent unethical practices	111
9.2	RICS codes of conduct	114
10.1	Rank order of job satisfiers for a range of construction personnel	119
11.1	Monitoring checklist	137
11.2	Example of a cost plan	141
11.3	Project handover/test certification checklist	144
11.4	Fire safety project completion checklist	145
11.5	Lessons learnt proforma	146

Foreword by Miles Wade

Having now committed to a major construction project, after a thorough planning process with one of the authors by my side, I feel well qualified to comment on the approach taken by the authors in this very helpful book. They clearly outline the nature of the risk v reward venture undertaken by a client, a designer and a contractor, and as *the client*, I readily accept that I have had to take the greatest risk but had the least knowledge of the culture and protocols of the construction industry.

The process began in earnest for me by a review of our strategy, setting out ten year goals, defined as 'the ends' which we were seeking to achieve, determining 'the ways' in which these goals were going to be delivered, and then ensuring that they were balanced with appropriate resources – 'the means' – over time and space. A key goal was to transform one of our sites by constructing new facilities which would broaden our appeal for more of our members and increase use of that site.

All of this required us to take on the role of a *'quasi-developer'*, not only developing the concept for the business idea, but also allocating the right resources and creating the environmental conditions for success. The approach taken was not to view it just as a building project where the architecture would create the market and draw attention to the new facilities, but to broaden the scope of the whole business venture through multiple connected 'soft' projects.

As a client, there needs to be a simultaneous learning of the design and construction process, as well as shaping the organisational thinking to initiate the project. The client has to be determined to avoid inefficiency and poor predictability of delivery against success factors of time, cost and quality in order to produce successful outcomes and ensure that the 'clients' requirements' were met.

In order to achieve the long term transformational change, it has been essential to obtain stakeholder engagement and buy-in, as poor delivery would reflect badly upon the competence of the organisation beyond the scope of the project – and risk damaging the reputation of the institution. Conversely, a project delivered well can significantly enhance the reputation of the organisation – and that is what we set out to achieve in following the processes described so eloquently in this book.

The authors identify that the greatest risk is *poor preparation*. They explain that the pre-project phase needs to start one to two years before an architect is even engaged. Poor preparation will have consequences upon each stage of project delivery thereafter. The client cannot just call upon a *'professional body of knowledge'*, yet they must transform to become a 'professional developer'.

Clients need to be equipped with the tools to lead their project. It requires an informed, methodical and informed approach – with clear roles and responsibilities. The methodology outlined in this book is a 'step by step' practical guide from the perspective of clients and their respective organisations. The book seeks to inform clients in a pragmatic way to establish a pattern of behaviour to de-risk a construction project. This optimum behaviour has been brought together into a single rational model, which the authors call the 'Construction Enterprise Risk Management' (CERM) model which parallels the RIBA Outline Plan Stage of Work, widely used by the designers and constructors throughout the industry.

The authors argue that, in the field of risk, there are the *Naïve* (that is, not recognising that one must go through a process of cultural change to deal with risk) to the *Novice* (that is, recognising that change is necessary without having the means to achieve it). This book aims to convert *Naïve* construction clients to become enquiring *Novices* – to enable a network of shared experiences and support.

This book also starts a debate about whether professional construction clients might learn better from conducting a strategic review which creates a vision at the outset, to optimising the asset in use many years after its handover. Construction literature habitually cast the client as a 'risk', when seen through the lens of designers and constructors, without fully understanding the client's point of view. The authors hope that, by defining the client's role in more explicit terms, it will enhance their support capability and reduce overall project risk. This should be welcomed by all sides of industry.

I hope this book inspires readers to take a different perspective on the procurement of construction services by clients and encourages transformational change in practice. I strongly commend it to those embarking or already engaged on transformational construction projects.

Brigadier Miles Wade CBE
Secretary of the Royal Automobile Club
Pall Mall and Woodcote Park

Foreword by Peter McDermott

The importance of client engagement in delivering successful project outcomes has been well documented in both research and policy publications in recent years. Welcome additions to the literature have been made that make a substantial contribution to understanding client engagement as a key part of the academic discipline of construction procurement and management. Having clients at the forefront of projects with all the necessary leadership skills, knowledge and resources has been highlighted as one of the main ingredients in improving the overall performance of the construction industry.

This improved knowledge and recognition has coincided in recent years with pressures on the performance of clients – market pressures on private clients and austerity measures on public clients. Maintaining the 'client voice' in briefing, decision making and approval processes, has become more challenging under these circumstances. If few clients play a proactive leadership role in the construction process, this may lead to inefficiency and poor predictability of delivery against success factors of time, cost and quality, never mind any requirements for the delivery of wider social value.

This dilemma around client engagement may present one of the contributory factors that has led to productivity in construction being consistently the slowest to change when compared to all UK manufacturing sectors. The current calls for transformational change through the Farmer Review and the Construction Sector Deal recognise the vital role of clients in helping to drive such change. This book helps make the case to clients and others that good changes to client relationships are not only necessary for good projects but have a significant role, in aggregate, as catalysts for transforming the industry.

This book will complement well the research, policy papers and generic client guides that are currently available. The book explores clients from many perspectives, and articulates the complex nature of their respective roles in the construction process. It delves into how motivational practices, ethics, governance consideration, selection processes for teams and leadership skills can have a significant influence on successful project delivery. Conversely, it articulates some of the reasons why projects fail and discusses lessons learnt from previous case studies as reflective practice for clients. Another issue considered in the book is the human side of construction management, especially since people embody client organisations and project teams. Organisation change, normally associated with major projects, can introduce emotion and uncertainty and needs to be sensitively managed by clients as a critical success factor accordingly.

The book has facilitated the creation of a practical guide, supported by the introduction of a 'client toolkit'. As part of the guide, a methodology is set down through process mapping, specially designed templates and proformas at each stage of the life of projects. This will allow clients to manage each and every stage and clearly communicate and define their requirements, in pursuit of meeting their business objectives. Furthermore, it provides innovative and unique initiatives for project managers to better achieve value for money and more effectively satisfy business objectives.

The book draws on case studies from the authors' experiences and interviews with construction practitioners. They address directly how clients can practically contribute to improvements to construction processes – throughout the supply chain. The authors place clients at the centre of construction management as project sponsors. This approach represents a unique, inventive and much welcomed pragmatic intervention to improving the industry. The book challenges readers to take a different perspective on the procurement of construction projects and services and guides practice that will encourage transformational change.

Written from the client perspective this book should sit comfortably alongside existing academic and client toolkits already on your bookshelves.

Professor Peter McDermott PhD
University of Salford
Joint Coordinator CIB W92 Construction Procurement Systems

Preface

Construction is a risk reward venture undertaken by a client, a designer and a contractor. However, it is the client that takes the greatest risk and, if it is a typical small enterprise, has the least knowledge of the culture and protocols of the construction industry.

Within the structure of a construction project, a series of government sponsored reports *Constructing the Team* (Latham, 1994), *Rethinking Construction* (Egan, 1998) and *Accelerating Change* (Egan, 2002) have made radical changes to the construction industry, making it more client focused than ever before. A greater sense of team working and integration between clients and the design and construction supply chain has now significantly reduced design and construction risk to, and from, the client. However, through the lens of designers and construction contractors the client can still be seen as a 'risk'.

Part of this risk can be characterised as the dynamic shift that a novice client has to undertake to become a *'Developer'*. The client is required not only to take on the role of delivering their own business and operational change but also to choose the right resources and create the right environment to successfully and seamlessly deliver a construction project. This is an enormous risk and time consuming to the organisation putting great strain on its management and physical resources. For many it is a 'leap into the dark' – the construction industry having few parallels in manufacturing.

The construction industry has been slow to embrace strategies linked to client leadership. Accordingly, the unique role of clients and their preparedness in projects is emerging as a 'hot' topic. This is not sufficiently covered by the professional institutions where the focus is on development of industry professionals rather than clients. Previous research within construction projects has mainly revolved around the development of professional teams which is well trodden ground and has increasingly diminishing returns on risk reduction.

To respond to this dilemma, the main focus of this book is to provide a suitable context for paradigm shifts in practice with measures to increase client involvement and leadership as catalysts for increasing the success of construction procurement strategies. It does not seek to fit clients within established project processes but focuses on where project processes and the standard models used by clients are able to co-exist and even become transferable skills to reduce project risk. The book is unique and outstanding in how it has responded in this dilemma in the following ways:

- It is designed as a practical 'how to do it' guide to the client role in construction projects.
- It has been created as a 'taster' with the intention of having a 'follow up' series of courses and continuing professional development. Furthermore, it is planned to build on the book's central proposition that the development of a client framework can reduce project risk.
- It is a comprehensive and wide ranging analysis of best client practice in the construction industry as well as other industries and sectors and what can be learned from them. The study will take established and widely accepted business management practices and models and align them with development/construction issues.

Delivered successfully, a construction project can enhance the reputation of an organisation. Delivered poorly, it reflects upon the competence of the organisation beyond the scope of the project itself.

The greatest risk is *poor preparation*. The client pre-project phase starts one to two years before an architect is engaged and it can have significant consequences upon each subsequent stage of project delivery. Yet, in the authors' opinion, most of the current literature for clients does not adequately prepare them for the whole 'journey' of developing a construction project from a business idea and managing the asset after handover.

The risk is that for any client searching guidance and direction they may have to read widely or seek help to understand their role. Clients are not the backdrop to a design story; the centrality of the business proposition, a strong 'client voice' and the 'critical success factors' and business 'red lines', emphasized in this book, are all essential to making the operating building a successful long term asset and change management story to the client and design team.

In the field of risk there are the *Naïve* (that is, not recognising that one must go through a process of cultural change to deal with risk) to the *Novice* (that is, recognising that change is necessary without having the means to achieve it). This book is aimed at least at converting *Naïve* construction clients to become enquiring *Novices*.

This book seeks to inform clients in a pragmatic way of their responsibilities in a construction project, but it also highlights the established pattern of behaviour that regular and successful construction clients follow to de-risk a construction project.

This best behaviour has been brought together into a single rational model, which we call the Construction Enterprise Risk Management (CERM) model which parallels the RIBA Outline Plan Stage of Work widely used by the designers and constructors that the client will find defines the construction industry's approach and language.

Part 1 of the book gives an interesting introduction as context into the construction industry and construction clients. It identifies some deficiencies and failures in terms of client involvement and leadership in achieving successful outcomes. To address the dilemmas around this problem, it also suggests measures to improve client practices through many different levels of intervention, resources and processes. Part 2 is deliberately designed as a handbook, a reference manual and practical 'how to do it' page by page guide, with further support from more practical guides and case studies on our website (www.innovateestates.co.uk). The authors hope to build a network of small businesses that share their stories and experiences and support each other.

This book is also to start a debate about the professional construction client and how they can best be supported in the unique set of skills that a construction project will demand of them. There is no body of knowledge to support clients in their duties to manage major construction projects or the potential corporate risk that can arise from starting a project.

The formulation of a vision from a business model to a fully fledged optimised asset in use many years after its handover is an experience most clients find tremendously satisfying and rewarding.

With the help of this book as a guide the scale of client involvement might be clearer from the outset. Success is not just a satisfying outcome, it is the smooth and orderly path through which it is achieved.

The research for the book has been derived from a combination of sources which include the authors' own experiences, coupled with interviews with a wide range of construction professionals and literature. The book is mainly intended for construction management practitioners and clients but could suit a wide target audience including under and postgraduate students and academics. The authors are hopeful that it will make a constructive and useful contribution to the field of client leadership in construction management.

References

Egan, J (1998). *Rethinking Construction: The Report of the Construction Task Force*. London: DETR.TSO. 18–20.

Egan, J (2002). *Accelerating Change: Rethinking Construction*. London: Strategic Forum for Construction.

Latham, M (1994). *Constructing the Team*. London: The Stationery Office.

Acknowledgements

The authors would like to thank all the research participants who contributed to the book findings through interviews. A special thanks goes out to Hawre Baban for his contribution to the book on critical success factors for construction clients in Chapter 6 and Shaun Taylor for the work around the importance of selection and appointment processes in Chapter 8. We would like to thank Stephen Caswell for his support. We are also indebted to the support of our wives and families during the long hours of preparation, thank you Jenny, Izzy, Esther, Ben, Margaret, Bobby and Kristin for your patience!

Part 1
An introduction into the construction industry and construction clients

Jason Challender

1 Introduction

> The owner of a project must provide clear direction and timely decisions, and must assist the project management team to drive the project to a successful conclusion.
>
> (Thompson, 1991)

It is perhaps the above quotation which has provided the focus for this book in an attempt to encourage clients to take a more proactive stance in project management, change current working practices in the construction industry and improve project outcomes. Accordingly, the main focus of the book is to explore the role of clients in construction management. In this regard, the overarching aim of the book is to create a factual client 'how to do it' guide or 'toolkit' for procuring more successful project outcomes. It is intended that this practical guide for clients can develop into a common due diligence framework on how to initiate, procure and manage construction projects and developments. From this perspective, it will raise awareness of best practice and instil improvements in construction contracting with clients at the epicentre of project teams. It will seek to address the significant institutional risk that lies in the lack of a clear and consistent approach to the client role in projects and guidelines. Such an approach will constitute a viable tool in ensuring effective, appropriate and successful interfaces of clients in pursuit of improvements to construction management. Furthermore, it is also intended to provide an important insight into the influence of construction clients in the success of construction projects and redevelopment programmes.

The book investigates the current arrangements that exist within the global construction industry, to create a more comprehensive understanding of the problems of client knowledge, interface and integration within project teams. It explores and analyses the overall commitment of organisations to encouraging client engagement in all construction stages which could be hindering the overall effectiveness of construction projects. This is intended to provide a suitable context for paradigm shifts in practice with measures to improve client interface and leadership as the catalyst for increasing project success.

A deficiency in appropriate and strong client leadership in the construction industry has been highly documented by authoritative sources over many years.

The book will seek to address this ongoing dilemma and act as a catalyst for improvements to the construction procurement processes. This is intended to encourage more successful team integration and collaborative ways of working between clients, their appointed consultants and the whole supply chain. This is a deliberate attempt to improve client project management practices, which have arguably not been delivering the impact and benefits in terms of successful collaborative project outcomes.

The book is intended to assist academics, construction related practitioners and clients in their awareness, breadth of knowledge and comprehension of the issues around client leadership, with the overarching aim of delivering projects that are more successful. This is felt to be particularly important as in previous studies into construction clients, very little attention has been focused on giving practical advice. The book has sought to infill the literature gaps through examination of traditional client roles and through providing guidance through the toolkit on potential improvement measures. Case studies and practical examples have been included to assist the reader on how theoretical perspectives can be applied to real life construction projects and scenarios. The book has also addressed academic calls for greater insight into construction client leadership that can be created, mobilised and developed and more understanding of the resultant positive effects and impact that can be generated therein (Walker, 2009). There will be frequent reference to construction practitioners' views and opinions throughout the book and these have been sought through research carried out in 2017 from a small sample of semi-structured interviews. Participants from these interviews included clients, design consultants, main contractors and subcontractors. Other findings have been sought from the widespread experience of the joint authors of this book.

There have been few books which have been written on the specific subject of incentivising appropriate client leadership in construction specifically through a 'client-side' practical guide or 'how to do it' toolkit. Those which have been published have largely focused on theoretical studies examining different client behaviours and relational analysis of clients with construction teams. Furthermore, the component elements of client leadership have been covered previously but there has been very little to articulate how these can be incorporated into construction procurement strategies. The book, drawing on case studies from the authors' experiences and interviews, takes a different approach to construction clients by asking some very fundamental questions:

- What is the importance and influence of client involvement and leadership in influencing more collaborative working and project team integration?
- What is the extent to which informed clients can influence the success of construction projects?
- How can client interventions be best embedded into procurement of projects?
- What constitutes best practice and what is the extent to which the client's role can influence the success of construction projects?

In consideration of the above questions the book's objectives are:

- To be the standard reference for business people in understanding projects and reducing construction based risk.
- To explain in straightforward terms with practice based examples where the '*client factor*' has an impact on the project and the main differences of business as usual and project practices.
- To use case studies to look at patterns of client behaviours, and how these affect successes, failures and key risks of projects as perceived by clients and other stakeholders.
- To use the above as the '*client voice*' to provide a client checklist, and reduce employers' risks on projects.
- To identify valued knowledge, skills, attitudes, behaviours and business practice that clients use in their approach to projects.
- To identify a set of clear guidelines, national or international, to support the client role.
- To form the basis of a practical toolkit for guidance and teaching around the unique role clients have in project development.
- To look to future developments and identify the key role clients take in Building Information Management (BIM), new developments in the RIBA Outline Plan Stage of Work and other areas such as continuing project integration and collaborative working.

The book is mainly intended for construction management practitioners and clients but could suit a wide target audience including under and postgraduate students and academics. For clients it will provide a summary of best practice and guidance in client project management from initiation to post-handover across many different client sectors. For professional practitioners, the book will explore how client management can be a complementary project skill and whether such skills can be proven to reduce risk in projects. The qualitative research for the book has consisted of interviews with a small sample of experienced construction clients. Accordingly it will provide the 'voice' of professional and institutional clients and present their recommendations and mechanisms for the creation of a common standardised client protocol. Furthermore, it will also consider whether this approach might complement other developments in project management such as project team integrated working and BIM.

The book's findings are presented to encourage professional practitioners to implement improvement measures through client intervention mechanisms and initiatives. The introduction of such mechanisms are explained in the book and presented as a practical guide, or toolkit, for improvements in construction project management practice. Reading this book will hopefully support the development of a deeper understanding of the benefits of having strong client leadership for improved outcomes for construction projects. With a better insight to how clients can be instrumental to project success it should provide the potential to embrace the true philosophy of collaborative working and therein promote better client management practice. The book is not intended as a holistic course textbook albeit it could be a worthy inclusion on a recommended reading list for courses related to construction procurement. The toolkit for improving clients' involvement on projects could be used as a basis for short term training or conference proceedings for professional institutions and public sector organisations. Notwithstanding this, it is not intended solely as a practitioner guide. Rather, the book aims to cross this divide and provide useful insight to both academics and practitioners in developing their understanding of the topic area.

Understanding the risks posed for clients in project management is a growing area as the tolerance for project failure reduces. Contemporary books to our own seek to externalise client issues, rooting their viewpoint within established project processes that see clients as an external stakeholder to be managed by project teams. This book takes a different perspective and is unique in considering the client as an integral and valuable member of the project team, instrumental for shaping the brief and subsequent design processes around organisational business cases. In doing so it creates a practical baseline for smoothing the transition between standard business and project practice and a starting point for standard business practice can be developed and integrated within project theory to support and de-risk subsequent project management.

The value proposition of this book is that it will be read, understood and accepted by business people as their main guidance and reference tool for reducing construction based risk. Accordingly, this is a book written for business people by business people based on sound theory (how to do it) and sound practice (lessons derived from case studies). It is considered unique in that it represents a comprehensive and wide ranging analysis of best client practice in the construction industry as well as other industries and sectors and what can be learned from them. The book will take established and widely accepted business management practices and models and align them with development/construction issues. From this perspective, it does not seek to fit clients within established project processes but use standard client business models to co-exist alongside such processes.

Although the research was undertaken in the UK, and all findings are likely to therefore have best fit with the UK construction industry, the overall knowledge and understanding to be provided by this book will have international relevance. Other countries seeking to develop client guides using similar approaches to the UK will be able to utilise the book, with consideration of how the findings fit with their own understanding in practice.

Finally, it is worth acknowledging that both authors have individually gained over 30 years' experience of construction management from both practitioner and academic perspectives. From this, the book has drawn on both academia and practice, and it seeks from both these perspectives to prove an important insight into an area which has long been problematic for the construction industry.

References

Thompson, P (1991). The client role in project management. *International Journal of Project Management*, 9, 90–92.

Walker, A (2009). *Project Management in Construction*. 5th ed. Oxford: Blackwell Publishing Ltd. 150–158.

2 A model of construction clients and their projects

2.1 Clients and the process of construction

This chapter of the book is largely related to the findings of Boyd and Chinyio (2006) and Kamara *et al.* (2002). When we refer to construction clients it is worth considering what we mean. In this regard, Kamara *et al.* (2002, p. 2) articulated that:

> A client can be defined as the person or organisation responsible for commissioning and paying for the design and construction of a facility (e.g. a building, road or bridge). And is usually (but not always) the owner of the facility being commissioned. The client can also be the user of a proposed facility, or they (i.e. the client and user) may be separate entities.

Notwithstanding the above, clients represent other interests, as the purchasers of services for the design and construction of a project. These are illustrated in Figure 2.1.

Such other interests could include other organisations affected by the projects being procured. Examples of these organisations may be funding bodies, neighbourhood organisations, local authorities and community groups. The degree to which these other organisations are affected by projects will depend on the nature and scale of developments which clients procure. For instance, a major regeneration project will carry far more influence on local organisations than a minor refurbishment or repurposing of an existing building.

Clients are important to the construction processes as they are normally the creators and funders of projects and as such the drivers for their developments. The ultimate goals of projects should be geared around the clients' requirements in terms of their aspirations, ambitions, visions, aims and objectives. Accordingly, it is of paramount importance that the requirements of clients are fully articulated and understood by themselves and others involved in their projects. This premise is justified on the basis that several projects have been completed in the past that have not been deemed successful as they have not fully encapsulated briefs set down by clients. For this reason, construction clients should, from the outset of projects, normally at the business case and conceptual stages, fully consult with their respective design teams and end users to expressly formulate their design

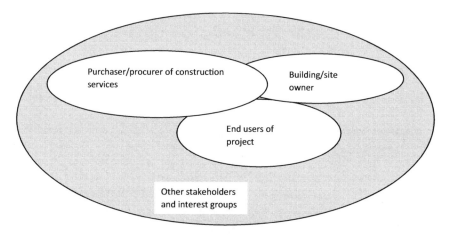

Figure 2.1 Integral components of the client body
Source: Adapted from Kamara et al. (2002).

briefs. From time to time, such briefs can change for many different reasons as the project environment changes or simply because clients need to rethink what they originally envisaged they require. One of the most important aspects for clients to control through the design and construction processes is variations or 'scope creep'. This can be particularly disruptive and expensive for projects if changes are made at relatively late stages in the programmes which could include abortive design works or main contractor claims for variations. Having a degree of clients active in all stages of projects as part of their roles and responsibilities is considered crucial for successful outcomes to be achieved. This will ultimately involve timely and proactive decision making at certain critical milestones in the life of projects. Any unnecessary procrastination in clients' decision making could result in project uncertainty and in some cases irreversible delays for projects or indeed programmes.

2.2 The 'client voice'

When we refer to the clients' voice, we are normally meaning a process by which their requirements in the development of building projects are systematically established and developed. This has been articulated in the quotation below:

> The 'voice of the client' (or clients requirements) includes the collective wishes, perspectives and expectations of the various components of the client body. These requirements describe the facility that will satisfy the client's objectives (or business needs). Client requirements constitute the primary source of information for a construction project and, therefore, are of vital importance to the successful planning and implementation of a project.
> (Kumara et al., 2002, p. 4)

In the past, the construction industry has been heavily criticised for not taking on board the involvement of construction clients in the processes of procuring projects and this has formed the basis of government reports aiming at improving professional practices (Latham 1994; Egan, 1998). Such reports called for more client orientated construction processes that involve clients at the heart of discussion and decision making. The practice, wherein professional consultants design buildings with the needs of environmental and appearance considerations in mind and not so much of those of the clients, needs to be carefully considered. Arguably, clients know their business requirements better than any other parties, and thus it is imperative for them to be consulted and listened to by their respective design teams, especially when discussing operational requirements and functionality. If those fundamental needs of clients are not met, it could have a detrimental effect not just on completed projects but on the longevity of clients' businesses. Accordingly, they should have a pivotal role in all stages of design and be able to appraise all decisions made by their teams to gain their approval or input. For all the above reasons, clients' involvement at the early stages of projects is certainly of paramount importance. In meeting clients' requirements, there is seldom any instances where only one proposal fulfils such needs. As such, it is common for many different alternatives to be considered from the outset and this could involve redevelopment of clients' existing premises or even moving their businesses elsewhere. To not consider all the options and alternatives could lead to the wrong clients' briefs being developed and not allow opportunities for optimum building solutions to be generated. This is normally referred to as 'value management' (Walker, 2009) and can be generated through business plans and improved value for money for clients.

2.3 The complexity of clients

The degree to which client organisations are 'complex' will vary according to the composition, size and nature of their respective organisations. They vary from being small family run businesses to large global organisations or public sector bodies. Notwithstanding this differentiation, the same process or *modus operandi* for considering and progressing projects should be appropriate and this relates to the 'business needs'. Such business needs could revolve around the requirements for additional space, more functionality of space, condition of premises or related to installation of new technologies. The overarching question of whether projects should progress at various stages of the development processes should be revisited in terms of business needs. Decisions around such organisational requirements are not always made by one individual but normally the organisation as a whole through project structures and boards. In large organisations, the structure of reporting and decision making through various tiers of hierarchy is complex. This is particularly the case with government, educational establishments and other public sector bodies. It is not unusual in these types of organisations, for decision making on major issues to go through a process of committee or board approvals prior to final approval at the highest executive or non-executive level. In some larger organisations, there can be conflicting requirements of what one sector or department wants over that of another. This could be particularly relevant where one part of an organisation is going to be significantly affected, positively or negatively, by a construction project or programme. Notwithstanding this premise, client organisations should consider and make decisions on the business needs as a whole. An example could be where a project is predicated on the needs of an organisation to rationalise space whilst improving communications and collaborations within its workforce. In such cases, an open plan office strategy might be considered for reasons of integration of staff and to facilitate efficiencies in accommodation. These collectively could contribute to meeting the business needs of the organisation. It may, however, not achieve the wishes of those staff who respect their privacy and do not relish open plan working. This reinforces where organisational requirements may be at odds with the needs of individuals.

Complexities can also arise from discrepancies between the different clients in terms of their respective degrees of 'experience'. Some clients may understand construction processes as they may be professionally qualified. Others may have some knowledge, owing to having prior experience of working with construction teams. Furthermore, there may be some who have no experience or knowledge of construction processes, which can sometimes present challenges for them in understanding the nature of the 'journey' through the design and procurement of projects. When this differentiation of both complexity and experience of clients is brought together, this can affect the requirements processing for projects, and this is illustrated in Figure 2.2.

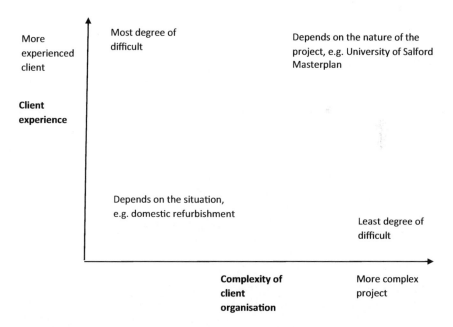

Figure 2.2 The influence of the clients' organisation and experience on information processing at briefing stages
Source: Adapted from Kamara *et al.* (2002).

What can be particularly challenging for projects is where very complex organisations have construction clients who are regarded as being relatively inexperienced. This represents a potentially difficult situation where the degree of risks on the project could be substantial as a consequence of the '*client voice*' not being properly established. The opposite is true at the other end of the continuum where organisations are not complex in their nature and where clients' experience is relatively high. In this latter case, there is much more scope for the client voice to come through in a clearly articulated and informed way. Notwithstanding this assertion, each project is bespoke and situation different for each, and much will rely on the experience and industry professionals involved including the design teams and contracting parties.

2.4 Client requirements through project briefs

The construction briefing process is the means by which clients can inform others of their aspirations, needs, ambitions, aims, visions and overarching desires for projects. According to Kamara *et al.* (2002) they should incorporate the following information:

- The background, purpose, scope, content and desired outcomes of the projects.
- The functions of the intended facility and the relationships between them.
- Cost and time targets, instructions on procurement and organisation of projects.
- Site and environmental conditions, safety, interested third parties and other factors that are likely to influence the design and construction of a facility.

When we consider capturing clients' requirements for projects, it is crucial to regard 'briefing' as an ongoing process, with the '*client's voice*' at the centre of consultation and decision making. If this is not the case, projects may become 'skewed' to meeting the requirements of other design consultants on the basis of their individual specialisms or preferences. This dilemma could become apparent where clients' requirements could become interpreted poorly by project teams, and the wrong brief emanates from this predicament.

Normally the briefing processes should follow staged and methodical approaches. In the UK, and in much of the developed world, the briefing process for design stages follows the RIBA (2013) stages and these will be illustrated and discussed more in the next chapter.

Ideally, each and every stage of the RIBA process should be signed off before moving to the next stage. This avoids the risk of progressing to the next stage of design and incurring abortive costs and time delays, before the preceding stage is fully understood by all and formally approved. This same methodology is also recommended as part of PRINCE2 principles for managing projects. Information should be made available to design teams by construction clients at the earliest stages of projects. Such information may comprise particulars or reports relating to the proposed site. Furthermore, information around the client's business should be used as a valuable source of data and this could be related to utilisation, contamination, asbestos register, site conditions, building surveys or record drawings. Should this information not be made available by construction clients early in the life of projects, delays associated with wasted time and resources could be experienced.

Briefing is a process which is linked to inputting information and transforming it in a creative way to produce an output (brief). The importance of consultation at the briefing stages, should ideally involve all those individuals as 'key stakeholders' that are considered relevant. Working with stakeholders should ideally encapsulate the following aspects:

- Identification of stakeholders
- Determination of what interest each stakeholder will have in the project
- Degree and extent of information each stakeholder will require
- The role of each stakeholder and decision making
- Communications strategies for stakeholders

One of the common mistakes that construction clients make is to not include those informed 'end users' in the early stages of consultation. This category of staff, within clients' organisations, are regarded as being 'key stakeholders' in the process of giving informed operational briefing information to enable designers to have a well grounded understanding of certain organisational aspects of companies. An example could be the engineering technicians when procuring a specialist workshop, and understanding from these individuals the functional and technical requirements for the bespoke space.

Information to form the basis of briefs should be collected from numerous methods and these could include the following:

- Workshops
- Examination of documentation provided by clients
- Inspection of existing buildings and facilities
- Electronic equipment for time/space auditing studies
- Visit to similar facilities that represent good examples of design and construction
- Interviews with key members of staff and other consultees
- Discussions with third parties who have a vested interest in the project, e.g. funders or local authority representatives

In compiling this information consideration should be given to the following aspects:

- Recording and compiling the briefing information
- Decision making in the briefing process
- The importance of value management at the briefing stage
- Considering and reflecting client priorities
- Processing of information, translation of client requirements
- Programme for implementation of projects

- Project tolerances and risks
- Budgetary/funding constraints

An example of a briefing proforma document that could be utilised as a client briefing checklist, and thereby reduce the risk that something could not be included, is illustrated in Table 2.1.

Table 2.1 Client brief checklist

About the project
☐ Project objectives and goals
☐ Business case for the project
☐ Location and key parameters
☐ Budget/financing
☐ Identification of key stakeholders/end users
Scope of required services
☐ Extent of services required from estates
☐ Identification of exclusions (e.g. FF&E)
Constraints and dependencies
☐ Any specific constraints on the scheme
☐ Key input required from other departments
☐ Key decisions on which the project is reliant
☐ Any client procedures/processes that need to be adhered to
Finance and procurement
☐ Any predetermined team or procurement route
☐ Investment background
☐ Funding requirements
Timescales
☐ Required occupation date
☐ Project spend date
☐ Phasing requirement
☐ Key milestone decision dates
☐ Key critical dependencies
Management and communications
☐ Proposed engagement plan
☐ Information management/communication plan

In terms of those individuals that should be part of the briefing process it could involve the following complement of design professionals:

- Project managers
- Architects
- Quantity surveyors
- Mechanical and electrical engineers

- Structural engineers
- CDM coordinators
- Landscape architects

In addition to the above, on particularly complex facilities it could be deemed necessary to include some specialist engineers. An example could be where a new concert hall is being designed and the need to appoint specialist acoustic/sound engineers at the early stages.

2.5 The transformational nature of client projects

2.5.1 Buildings, organisational and human considerations

According to Boyd and Chinyio (2006), the basic structure of a construction client model is internally made up of individuals, the organisation and building. This can be illustrated in Figure 2.3.

This model is based on a 'systems dynamics model' as advocated by Flood and Carson (1993). In its simplest form, it contains three areas which construction clients need to feel satisfied with; namely buildings, organisational considerations and people. There is a means-ends dimension which involves change and time, and can be deemed to be what clients experience. There is also a processes and knowledge continuum, which takes into account decision making, communications and conception. This involves how clients see, make

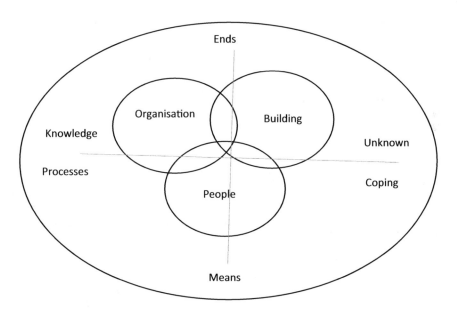

Figure 2.3 Basic structure of a construction client model
Source: Adapted from Boyd and Chinyio (2006).

decisions, communicate information and perceive things. From these, it follows that projects need to be perceived as successful from buildings, organisational and people perspectives. The normal goals for the success of projects relate to buildings, but in reality there are many other organisational factors. These include the business objectives, which could be part of the corporate and organisational strategies, and how developments are funded. Accordingly, achieving organisational satisfaction is normally regarded as far more challenging than purely procuring successful buildings. When considering people, the client body is normally made up of many individuals within their respective organisations. With regard to this diversity of workforce, client staff expectations and needs require to be considered and satisfied. Sometimes this process of meeting staff requirements can induce emotions, especially where they are impacted directly by projects. Examples could be where staff are temporarily or permanently displaced as a consequence of new buildings or adversely affected by disruption from the construction works. Human needs or expectations can be different, depending on the variety and type of individuals which adds further challenges for construction clients in meeting their requirements.

2.5.2 Organisational changes in clients brought about through projects

When undertaking building projects, it is clearly apparent that there will be changes to physical environments. What is less understood or clear is the change in the culture or structure of client organisations brought about by some building projects. This is especially the case where transformational change is a corporate objective, and physical development of buildings or the client's estate is regarded as the vehicle for driving change. Such organisation changes normally commence prior to the construction phases of projects and continue after building completion. The organisational changes are normally instigated by clients but not all clients understand the importance of properly managing the change processes. Clients, for instance, may want to design their new buildings in a way to encourage organisation change in the form of improvements to human and cultural aspects of their businesses. Examples could be where offices are designed in an open plan configuration, to promote human interface and improve communications amongst the workforce. Interventions such as these, have in the past been proven to be successful but the mode in which the changes are brought about with sensitivities around emotional consequences and anxieties must be fully understood by construction clients. In this regard, successful strategies have revolved around involving staff in discussions around reshaping their working patterns, and consulting them at each and every stage of the change processes. This ensures a degree of 'buy in' where staff can be encouraged to promote such initiatives. Notwithstanding this premise, the adverse effect could ensue if staff are not involved or consulted. In these cases, staff within client organisations may feel that changes have been imposed on them without any of their concerns, aspirations or ideas being considered. This can nurture a culture of indifference to the changes and be self-defeating for clients in terms of improving attitudes, behaviours and working practices. Accordingly, construction clients should adopt democratic strategies linked to obtaining individual and group consultation and being perceived to be acting on feedback and contributions that come from staff invention.

Organisation changes brought about by projects may be linked to company desires to contract, expand or conduct their business operations differently, possibly as a response to changes in the business environments they work within, and staying ahead of the competition. Such changes could also be a response to technological advancements that organisations wish to introduce or simply to change the perceptions of their respective organisations. This could be particularly relevant where clients are operating within old buildings. In this scenario, functionality to meet their operational requirements may present an issue for clients. In addition their accommodation may not present a good representation or perception of the brand that the client organisation wants from the marketplace.

2.6 Project and client requirements

2.6.1 *Understanding the businesses of construction clients*

According to Boyd and Chinyio (2006), in order to understand client businesses and their operational requirements, it is necessary to know the corporate environments that they work within. This involves examination of the purpose or service of their respective organisations, together with their company structures and the defining processes of their businesses. This can entail researching aspects of organisations, through known information and published data, or alternatively through consultation with many different individuals within those organisations. Furthermore, it could include conducting interviews with key members of the clients' management teams to determine how they see their organisation responding to certain situations and eventualities. This could serve to provide a holistic insight into businesses rather than simply a functional view of their operations. In this pursuit, it may be useful to consider the historical perspective of organisations, as past events may have a bearing on their direction of travel. For instance, past failures and lessons learnt from previous ventures will not want to be repeated, whereas their corporate *modus operandi* might be to build upon their past successes as a recipe for meeting business aims and objectives.

According to Boyd and Chinyio (2006), there will be internal differentiation and external influences on construction clients' businesses. The diagram in Figure 2.4 illustrates this analogy with core business at the centre and agents and stakeholders at varying distances from the centre, dependent on their respective degree of influence on clients' businesses. Those near the centre would be integral to the decision making and influences on strategic planning whereas those on the periphery would be more related to the external environment that organisations work within.

It is important to have an appreciation of the effect and influence that any proposed building project will have on any client organisation. This may

18 *The construction industry and clients*

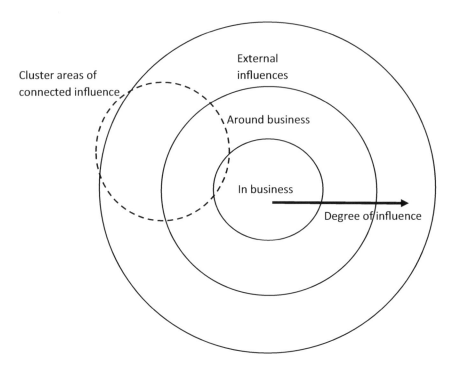

Figure 2.4 The client's business operation model with influence mapping
Source: Adapted from Boyd and Chinyio (2006).

be specifically set down in business plans or strategic plans adopted by those businesses. In addition, it is critical to identify what are the client organisations' internal drivers including overarching visions, aims and objectives. These could vary between short, medium and long term. The diagram in Figure 2.5 illustrates a 'means and end' map which could involve internal and external influences. Those internal drivers would normally revolve around operational, tactical and strategic aspects whereas external drivers would normally be related to operational, managerial and political issues. It is of paramount importance for construction clients to understand these influences, as otherwise their ability to lead projects to successful outcomes may be compromised.

2.7 Summary

This chapter has articulated how construction client organisations are diverse and come in many 'shapes and sizes'. Examples of these organisations were given as funding bodies, neighbourhood organisations, local authorities and community groups. Furthermore, they may comprise small family businesses or charities. The ultimate goals of projects should be geared around the clients' requirements in terms of their aspirations, ambitions, visions, aims and objectives.

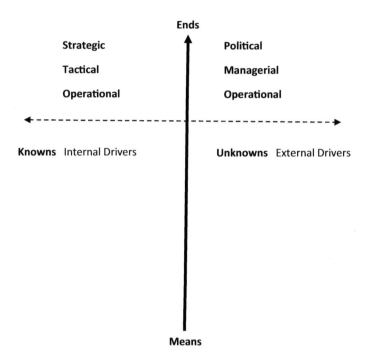

Figure 2.5 Internal and external influences on a client organisation
Source: Adapted from Boyd and Chinyio (2006) and Kamara *et al.* (2002).

Construction clients should from the outset of projects, normally at the business case and conceptual stages, fully consult with their respective design teams and end users to expressly formulate their design briefs. In formulating briefs and directing their project teams, it is important for the '*voice of the client*' (or clients' requirements) to be clear and unambiguous. Communication from clients in this regard should include the collective wishes, perspectives and expectations of the various components of the client body. These requirements must describe the means and vehicles by which the client's objectives (or business needs) will be satisfied. One of the considerations in this regard is to ensure that construction clients have a pivotal role in all stages of design and be able to appraise all decisions made.

Each client body can be differentiated on their organisational complexity. In this regard, the degree to which client organisations are 'complex' will vary according to the composition, size and nature of their respective organisations. They vary from being small local businesses to large global organisations or public sector bodies. Complexities can also arise from discrepancies between the different clients in terms of their respective degrees of 'experience'. What can be particularly challenging for projects is where very complex organisations have construction clients who are regarded as being relatively inexperienced. In such scenarios, this can add risks to projects especially if the client voice does not reflect what the client organisation requires in terms of business needs.

When we consider capturing clients' requirements for projects, it is crucial to regard 'briefing' as an ongoing process, with the '*client's voice*' at the centre of consultation and decision making. The importance of consultation at the briefing stages, should ideally involve all those individuals as 'key stakeholders' that are considered relevant. One of the common mistakes that construction clients make is to not include those informed 'end users' in the early stages of consultation and construction. Accordingly, they should ensure that end users are engaged throughout the whole life of projects, including post-completion. Furthermore, it is important for clients to have an appreciation of the effect and influence that any proposed building project will have on their organisation especially since there will be buildings, organisational and human considerations. This may be specifically set down in business plans or strategic plans adopted by those businesses.

References

Boyd, D and Chinyio, E (2006). *Understanding the Construction Client*. Oxford: Blackwell Publishing Ltd.

Egan, J (1998). *Rethinking Construction: The Report of the Construction Task Force*. London: DETR.TSO. 18–20.

Flood, RL and Carson, ER (1993). *Dealing with Complexity: Introduction to the Theory and Application of Systems Science*. New York: Plenum Press.

Kamara, JM, Anumba, CJ and Evbuomwan, NFO (2002). *Capturing Client Requirements in Construction Projects*. London: Thomas Telford Publishing.

Latham, M (1994). *Constructing the Team*. London: The Stationery Office.

RIBA (2013). *RIBA Plan of Work 2013*. Available at: www.ribaplanofwork.com [Accessed 23rd January 2018].

Walker, A (2009). *Project Management in Construction*. 5th ed. Oxford: Blackwell Publishing Ltd. 150–158.

3 The unique role of clients in the context of the construction industry

3.1 The bespoke nature of the construction industry

> The construction industry is considered to be one of the most dynamic and complex environments as it is a project-based industry within which individual projects are usually built to clients' needs and specifications.
>
> (Tabassi et al., 2011)

In considering the above quotation, it is interesting to explore from where the construction industry has emerged as a bespoke project based industry and there are many different theories around this. As a starting point, it is important to understand what makes the construction industry different and potentially 'at odds' from most other industries. In answering this question it is worth contemplating that the procurement process around construction is very unlike that of most other industries. Those employed are made up of mostly small teams, ranging from construction workers to design consultants, who come together on a temporary basis for the life of a project and then disband to undertake different projects. This creates fragmentation and does not always allow the time for relationships to develop and flourish, which could be in itself a contributory challenge for trust generation. It is also important to reflect upon the 'end product' and that construction projects are nearly always bespoke to clients' requirements. This 'one off' or 'made to measure' aspect does, however, create risk and uncertainty for all parties. To fully appreciate and understand this context it may be useful to compare the procurement of a new building with the purchase of a new car. When one buys a car the make and model that suits your budget will be agreed alongside any affordable optional extras that are required. One can even 'test drive' the same model to ensure that it meets the customers' expectations in terms of feel and drivability. At this stage, on ordering the vehicle, the customer will know exactly what they will receive on the due delivery date, which is normally a few weeks at most and have an agreed fixed price. In this regard, there is very little risk that the customer will not receive exactly what they have expected when they ordered for the price they have secured. As the car is made in a factory it will be standardised and quality control is normally very good accordingly. The complete opposite scenario could be argued to prevail when a building is being procured. It normally

Figure 3.1 The iron triangle of cost, time, quality and scope

involves a prolonged period of time for design consultants to formulate a brief with clients, progress the design development and tender the projects to construction contractors. On receipt of tenders, this is where the process probably varies most from the car purchase example. In selecting the most appropriate tender it is important to consider the quality of the tenders rather than just accept the lowest price. Such factors as reputation, track record, resources and demonstration of an understanding of the project are vitally important as the quality, cost, scope and timescale in delivering the final project are normally anything but assured. There are three elements which are widely regarded as critical success factors namely cost, time and quality and these form what is commonly referred to as the 'iron triangle' which is illustrated in Figure 3.1.

There are several unknown factors in any construction process which could cause the cost of projects to increase, programmes to be delayed and the quality of builds to be compromised. This introduces two aspects which come into play around commerciality and risk and who incurs any additional costs is frequently an area where disputes arise between clients and their consultants and contractors. Furthermore, given that construction works normally incur significant amounts of money the stakes are high in terms of the final bill for clients and the level of profit attained by contractors. It is therefore perhaps not surprising, for reasons of commerciality, that parties to construction contracts have traditionally not relied on trust in dealings with each other especially around financial construction matters.

3.2 The construction industry and recent reforms

The UK construction industry has arguably been associated, over many years, with projects which have had less than successful outcomes. This has culminated from many different reasons, and many measures and recommendations have come forward to address this ongoing dilemma and act as a catalyst for improvements. Some recommendations for best practice and obtaining more cost and time predictability on projects include measures to improve partnering procurement processes and therein encourage more successful team integration and collaborative ways of working. This is a deliberate attempt to improve construction practices, which have arguably not been delivering the impact, and benefits that were intended in terms of successful collaborative project outcomes. Various government reports have reinforced this dilemma over the years. These have included Latham (1994), *Constructing the Team*, Egan (1998), *Rethinking Construction*, and Egan (2002), *Accelerating Change*. Other examples include *Construction 2025 – Industry Strategy: Government and Industry in Partnership* (HM Government, 2013) which identifies that fractious qualities are embedded in the UK construction industry. In the latter case, the report stresses that collaboration and trust across the entire supply chain are crucial to deliver successful projects. More recently, this dilemma has been reiterated and become a major feature of *Modernise or Die: The Farmer Review of the UK Construction Labour Market* (Farmer, 2016). Furthermore, findings from *Low Carbon Construction Final Report* (HM Government, 2010) confirm the growing need for increased collaboration and integration across the industry, especially between the supply chain and clients, in order to make greater contributions to the pursuit of efficiencies.

3.3 The involvement of clients in the procurement of design and construction services

3.3.1 *Traditional procurement of construction projects: barriers and problems*

There is an argument that mistrust has been inherent within the UK construction industry for a long time between all parties including clients and the consultants and contractors they employ. It is important to comprehend the factors which build trust between clients, consultants and contractors in this regard. In this way, the critical factors can be realised and clients are then more able to facilitate alignment of organisational interventions to build trust (Hawke, 1994). This is especially important as the development of trust has proved problematical with only limited success owing to the different contractual interests of those involved (Lu and Yan, 2007). Maurer (2010) reiterated this argument but also concluded that retaining trust, in addition to building it, can be an equally challenging task. This is an area of study which has received only limited attention within the realms of construction management. According to Maurer (2010), in addressing these challenges, further research is recommended into different factors which influence the development of trust and likely outcomes and the book will explore the issues further in later chapters.

3.3.2 Deficiencies with traditional construction procurement

In considering the problem of trust between construction clients and those construction related organisations they employ, it is imperative to consider the traditional adversarial nature of the construction industry which has often been attributed to creating barriers for trusting relationships to grow. Most practitioners would argue that this has stemmed from the traditional forms of procuring construction work which over recent years have been blamed for achieving low client satisfaction levels, poor cost predictability and time certainty. Such a dilemma has largely been attributable to coordination difficulties associated with separation of design and construction and the greater need for teamwork (Latham, 1994, pp. 81–83; Egan, 1998, pp. 18–21; Egan, 2002, p. 6). Traditional or conventional procurement methods in this sense have been regarded as the standard practice in the construction sector for many years (JCT, 2014) and normally rely on completion of design and full documentation before tender. It has been predominantly geared around stages of construction procurement which include feasibility, design, tender, construction, commissioning and handover. Traditionally contracts have in the past normally been procured with minimal if any contractor, subcontractor or supplier design input in the early stages (MacKenzie and Tuckwood, 2012). This separation of design and construction is illustrated in Figure 3.2 which shows a structure chart wherein all of the design team are employed by clients at all stages of the project. Furthermore, Figure 3.3 illustrates under this traditional procurement route that all design stages of the RIBA Plan of Work 2013 are completed without any contractors' design or input.

Such contracts have been criticised for being awarded on lowest price tenders and having 'win' 'and 'lose' outcomes in the past where one party has overcome the other, normally in terms of commercial gains. It has also encouraged in some instances a 'blame game' when problems arise, most predominantly in the construction phase, through general lack of teamwork. This approach over many years seems rather fruitless as it has fuelled a culture and environment of mistrust between contracting parties to the extent that they are hesitant or unwilling to rely and vest trust in the other. From the clients' perspective this may manifest itself in feeling that contractors are 'out for what they can get' with maximisation of profits as the main motivator. Some clients and consultants often refer to contractors as 'claims conscious' in this regard. From the contractors' perspective, a lack of trust in traditional procurement may stem from them being responsible for construction of a project in which they have not had any involvement in any of the design, since they were invited to tender on full designs. As a result they may not approve fully of some of the design elements of the project and despite feeling that they could have made improvements had they been involved earlier albeit are ultimately responsible for delivering the completed

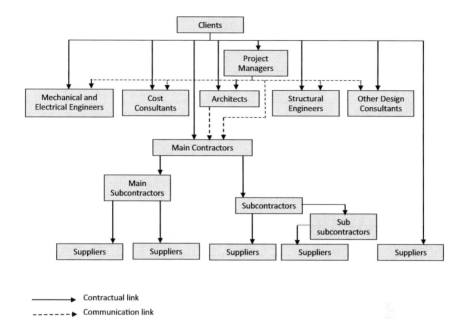

Figure 3.2 Diagram illustrating traditional procurement structure

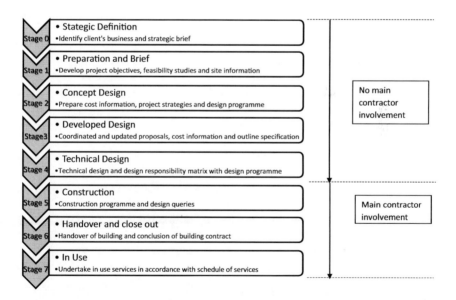

Figure 3.3 Traditional procurement route under RIBA Plan of Work

building. For these reasons, some would explain the lack of trust from the view and general perception that an embedded mentality is prevalent within the construction industry which relates to a general lack of integration of design and construction (Cartlidge, 2004, p. 11). The choice of procurement strategies on projects has therefore become a long contentious issue within the construction industry to address such perspectives. Other criticisms of traditional procurement methods have focused on their adversarial nature, deficiencies of design and construction interface and the inefficiencies that these can bring for construction projects. Recommendations for changes, designed as improvement measures, emanated from such criticism and included bridging the gap between design and construction and encouraging early contractors' involvement in value management and buildability (Emmerson, 1962; Banwell, 1964; Latham, 1994, pp. 40–51; Egan, 1998; Egan, 2002). Furthermore, others have outlined that risk management under such traditional arrangements is problematic especially when complex projects create greater risks for project teams. Contracting parties in these cases may seek to preserve their own individual commercial positions, frequently causing disputes to arise. These have in some cases led to reports of programme delays, cost overruns, conflict, distrust and legal action (Chan et al., 2008). To prevent such occurrences arising there is an argument that 'collaborative working' or 'partnering' offers a more suitable context for developing cooperation and trust as there is less focus on price and authority (Eriksson and Lann, 2007).

3.4 Overall context of collaborative working and partnering within the construction industry

The aforementioned perspectives would tend to suggest that traditional procurement strategies are widely considered to be inefficient and perhaps confrontational. Collaborative working at an early stage between clients and their design teams has been, post-Latham, regarded as a means to bridge the gap between design and construction to improve project outcomes. Accordingly, many have identified collaborative procurement routes through 'integrated' teams as a critical success factor on construction projects (Vaaland, 2004). Such procurement routes allow construction consultant and contracting teams to interact at earlier stages in the construction cycle by allowing the overlap of design and construction. This integration of design and construction is illustrated in Figure 3.4 with reference to a design and build procurement structure wherein the contractor employs the design team directly post-novation. Furthermore Figure 3.5 illustrates under this integration of early contractors' involvement into the design and build procurement route through reference to the RIBA Plan of Work 2013. These have been previously heralded as the means to address calls for change and fulfil the future challenges ahead but only if parties are prepared to build trust within such relationships (Kaluarachi and Jones, 2007).

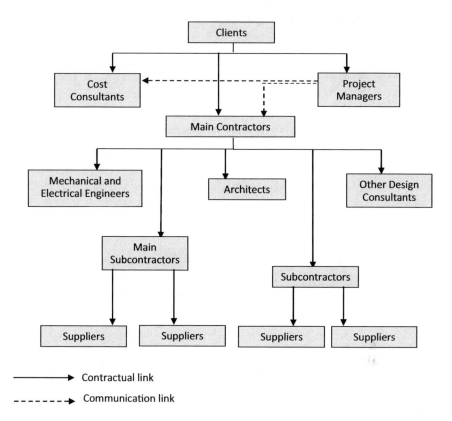

Figure 3.4 Design and build procurement structure

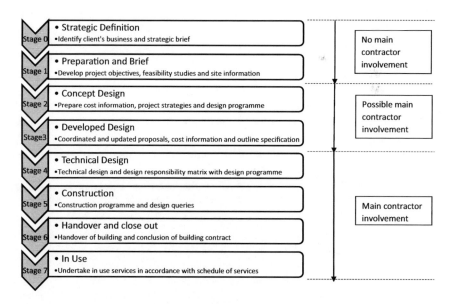

Figure 3.5 Design and build procurement route under RIBA Plan of Work

3.5 Alternative procurement methods for clients on construction projects

3.5.1 Government context: reports, codes of practice and recommendations for change

Throughout the decades there has been much criticism directed at the UK construction industry and many government reports have been commissioned over the years. Early reports in this respect included the Simon Report (1944), Emmerson Report (1962), Banwell Report (1964) and Potts Report (1967). These reports highlighted the deficiencies with construction related business approaches in terms of low performance, failure to meet client satisfaction levels and need for efficiency measures (Simon, 1944; Emmerson, 1962; Banwell, 1964; Potts, 1967). Perhaps, though, the most prominent report which has made recommendations on procurement in recent years is 'Constructing the Team' by Sir Michael Latham (1994). In this report Latham was critical of traditional procurement and contractual routes owing largely to a lack of coordination and integration between construction and design. He explained that in normal circumstances the main objectives of clients would include good value for money, attractive appearance, no defects, completion on time, fit for purpose, issue of guarantees and low maintenance costs. Furthermore, he described, owing to fundamental problems of conflict between employers and contractors, why these objectives have not been realised to their full extent. Accordingly, he suggested that a change of behaviour towards partnering was required to increase fairness, encourage teamwork and enhance performance through clients and consultants engaging more with contractors and their supply chain (Latham, 1994, pp. 50–57). The Egan Reports, *Rethinking Construction* and *Accelerating Change* (Egan, 1998; Egan, 2002) followed on from this, calling for integration of the design and construction phases with project teams working more collaboratively. These reports referred to construction projects in the USA where modularisation, standardisation and supplier/subcontractor design innovation had facilitated greater efficiencies than those procured in the UK (MacKenzie and Tuckwood, 2012). Figure 3.6 illustrates some of the perceived benefits, especially on complex projects that could emanate from the collaboration and early involvement of contractors, from findings of Latham and Egan.

To reinforce Latham's recommendations, the Construction Industry Council Strategic Forum for Construction (CIC, 2002, p. 15) reported that projects which had applied principles of both Latham and Egan in the use of collaborative procurement methods, have led to significant improvements in client satisfaction, cost predictability, safety and time predictability. Despite this, however, Constructing Excellence (2013) reported that there has been 'patchy' take-up of recommendations for adoption of collaborative working practices, mostly attributable to strict and inflexible UK public sector procurement rules.

The government study, *Construction 2025 – Industry Strategy: Government and Industry in Partnership* (HM Government, 2013), has identified low levels of

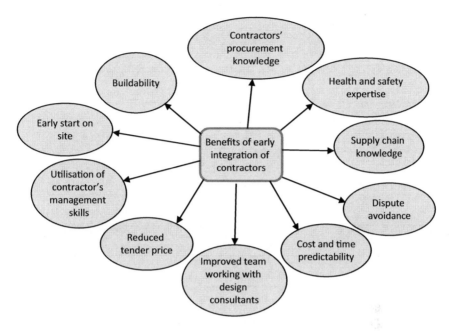

Figure 3.6 Benefits for clients in early integration of contractors

innovation and investment in research and development coupled with uncertain demand, which it claims is hindering collaborative working and partnering in the UK construction industry. It has also highlighted high construction costs as a major weakness in UK construction, mostly attributable to it 'being driven by inefficient procurement and processes'. For this reason, it states government's intention to lead the way in strategic procurement and sets down targets through both the *Infrastructure Cost Review* (HM Treasury, 2010) and the *National Infrastructure Plan 2013* (HM Treasury, 2013). Such targets reflect a reduction of between 15 per cent and 20 per cent on construction costs over a five year period. A further government report: *Low Carbon Construction Final Report* (HM Government, 2010), reinforced that partnering with extensive integration and collaboration within project teams remains a crucial vehicle to deliver successful construction projects. Key performance indicators linked to success, in this report, included achieving lower carbon emissions and improving predictability in terms of time and cost of projects through collaboration. The report also claimed that a downward trend in partnering has led to limited knowledge sharing, familiarisation and efficiencies being realised. In order to counteract this problem, it recommended greater collaborative working between different organisations including clients and their supply chain partners as the catalyst for reversing the trend. Further support came from the *Infrastructure Cost Review* (HM Treasury, 2010) which reported that procurement strategies particularly in the public sector have not always been efficient. It opined that 'public sector clients are more risk averse to the cost and time implications of potential challenges, and processes are overly complex and too much of a box ticking exercise'. This once again reinforces the need to reappraise procurement strategies and justifies the focus for strategies linked to increasing collaboration and trust.

The other major initiative that the government has been spearheading recently is the wider use of Building Information Modelling (BIM) and its intention to require that 3D BIM is mandatory on all public sector projects since 2016. This is part of the UK Government's four year programme to modernise the industry with key objectives to reduce capital costs and lower carbon emissions from the construction and operation of new buildings by 20 per cent (BIM Task Group, 2013). There are already, however, concerns being raised about collaboration and integration of the whole supply chain which is arguably one of the most important prerequisites to making BIM work effectively. These relate to the notion that main contractors, subcontractors and suppliers may not have an appetite for investing upfront resources in BIM, spending additional time on tenders and sharing their best ideas, if they perceive they only stand a one in six chance of winning contracts. This reflects tendering in the UK construction industry, traditionally geared around competitive bids from a number of potential contenders. However, it is unlikely that a BIM tendering process based on negotiation or fewer tenders would satisfy the market in terms of demonstrating competition. This is especially the case for public sector procurement with strict regulation and governance around such processes. Perhaps therefore a cultural change, as advocated by *Construction 2025* (HM Government, 2013), following on from Latham and Egan, might be the only means to reduce the UK's reliance on the ethos of competition.

In the pursuit of the government's drive to modernise the UK construction industry it has outlined that central to aforementioned challenges of reducing build and operating costs, the sector will need to adopt information technologies and reform collaborative behaviours and processes to realise more efficient ways of working. To assist in these reforms, the British Standards Institute has published a new standard *PAS 1192-2: Information Management for the Capital Delivery Phase of Construction Projects* with the aim of adopting a best practice and consistent approach around such initiatives (BSI, 2014).

3.5.2 Industry and professional context

In considering the problems around collaboration and trust, it is necessary for clients to understand the context within which the UK construction industry operates. On reflection, the environment is unique and arguably more complex than most other sectors which include manufacturing, as projects tend to be bespoke and require creative thinking and innovation at most times (Walker, 2009).

The other contributing factor which makes the construction industry unique is the fragmented nature of both the professional disciples and workforce within it. Donaghy (2009, as cited in Arewa, 2014) claimed that 192,000 independent small contractors operated in the UK in 2007, of which over 93 per cent had fewer than

14 employees. More recently official figures published in the Business Population Estimates 2014 (BIS, 2014) reported that there were 950,000 small and medium enterprises (SMEs) operating in the UK construction sector at the start of 2014. Fragmentation will emanate from such a large number of relatively disparate and mostly small specialist contractors and suppliers having 'arm's length' relationships on building projects (Briscoe *et al.*, 2001). This could hinder innovation and continual improvement from lessons learnt and experience with partnering being heralded as the means to address such a deficiency (Ikechukwu and Kapogiannis, 2014).

In consideration of the aforementioned difficulties there have been many reports of good practice and benefits arising from partnering (Rahman and Kumaraswamy, 2004; Walker, 2009). Unfortunately, not all organisations agree with such arguments and partnering has attracted its critics in some instances. RICS (2005, p. 2), for example, explained that recently reported successful experiences in collaborative procurement 'are largely anecdotal and attributable to the experiences of exemplar organisations'. It argued that the focus on success rather than failure has created an unbalanced view and presented a false impression in terms of the contributions made by partnering. Accordingly, it questioned the reliability and validity of some information sources. There is also an argument that partnering is losing popularity in a climate emerging from recession where some clients still view competitive tendering as the best means to secure best value (RICS, 2012). Furthermore, Morgan (2009, p. 9) concluded that with major capital projects, procurement routes that promote alliances and partnerships are not always appropriate and open to abuse owing to the scale of the commercial interests involved. This reinforces the previous arguments that partnering could be losing its popularity.

Notwithstanding these opposing views to partnering approaches, studies conducted by the Construction Products Association (HM Government, 2010) concluded that greater integration with supply chain partners could make a significant contribution to the low carbon agenda and reduce construction costs. Crespin-Mazet and Portier (2010) supported this and explained that the construction industry could benefit from a 'joined up approach' as already experienced in other sectors. This has served to 'front load' problem solving and lessen expensive design modifications later in the process. Furthermore, the British Standards Institution has recently developed with industry and introduced BS11000 which provides a framework for building and maintaining collaborative relationships between organisations. It is a model based on eight phases and designed to 'enable organisations to focus their efforts from concept to disengagement' (BIS, 2014). Initiatives such as these could be specifically designed to 'turn the tide' in favour of more participation in partnering practices. Perhaps these could make an impact and encourage more collaborative working in the construction industry.

3.6 Summary

This chapter has focused on the bespoke nature of the construction industry and the need for cultural and behavioural reforms to improve the relationship between construction clients and their contracting partners. This is aimed at improving performance for the industry and obtaining more successful outcomes for construction clients.

The construction industry has emerged as a bespoke project based industry where there are many different characteristics to other industries such as manufacturing. This is largely due to construction projects being nearly always unique and 'tailor made' to suit clients' individual requirements.

The UK construction industry has arguably been associated, over many years, with projects which have had less than successful outcomes. There are many different reasons that have emerged as to why this might be the case. Some relate to the short lived duration of construction projects whilst others relate to lack of continuity and fragmentation across supply chains. Various government reports have reinforced this dilemma over the years and identified that fractious qualities are embedded in the UK construction industry. Arguments have emerged that mistrust has been inherent within the UK construction industry for a long time between all parties including clients and contractors they employ. This may have stemmed from the traditional forms of procuring construction work which over recent years have been blamed for achieving low client satisfaction levels, poor cost predictability and time certainty.

To prevent such occurrences arising, there is an argument that 'collaborative working' or 'partnering' offers a more suitable alternative for construction clients in procuring more successful outcomes for projects. Projects which had applied principles of both Latham and Egan in the use of collaborative procurement methods, have led to significant improvements in client satisfaction, cost predictability, safety and time predictability. In this context, such partnering based approach have been proven to encourage cooperation and trust. These factors have been discussed in this chapter with a focus on client/contractor trust as a collaborative necessity, which construction clients should be encouraging and closely managing on their projects.

References

Arewa, AO (2014). *An Empirical Analysis of Commitment to Health and Safety on the Profitability of UK Construction SMEs*. PhD thesis submitted to University of Bolton, UK.

Banwell, H (1964). *The Placing and Management of Contracts for Building and Civil Engineering Work*. London: HMSO.

BIM Task Group (2013). *Welcome to the BIM Task Group Website*. Available at: www.bimtaskgroup.org/ [Accessed 17th December 2014].

BIS (2014). *Department of Business, Innovation and Skills: Business Population Estimates for the UK and Regions 2014*. Available at: www.gov.uk/government/uploads/system/uploads/attachment_data/file/377934/bpe_2014_statistical_release.pdf [Accessed 21st March 2015].

Briscoe, G, Dainty, ARG and Millet, S (2001). Construction supply chain partnerships: skills, knowledge and attitudinal requirements. *European Journal of Purchasing and Supply Management*, 7(2): 243–255.

BSI (2014). The British Standards Institution. *New Standard for BIM to Help Meet 2016 Government Saving Target*. Available at: www.bsigroup.com/en-GB/about-bsi/media-centre/press-releases/2013/3/new-standard-for-bim-to-help-meet-2016-government-savings-target/ [Accessed 17th December 2014].

Cartlidge, D (2004). *Procurement of Built Assets*. London: Buttersworth-Heinemann.

Chan, APC, Chan, DWM, Fan, LCN, Lam, PTI and Yeung, JFY (2008). Achieving partnering success through an incentive agreement: lessons learned from an underground railway extension project in Hong Kong. *Journal of Management in Engineering*, 24(3): 128–137.

CIC (2002). *Construction Industry Council Strategic Forum for Construction: Accelerating Change*. London: Construction Industry Council.

Constructing Excellence (2013). *Constructing the Team* (The Latham Report). Available at: constructingexcellence.org.uk/resources/constructing-the-team-the-latham-report/ [Accessed 18th March 2015].

Crespin-Mazet, F and Portier, P (2010). The reluctance of construction purchasers towards project partnering. *Journal of Purchasing and Supply Management*, 16(2010): 230–238.

Egan, J (1998). *Rethinking Construction: The Report of the Construction Task Force*. London: DETR.TSO. 18–20.

Egan, J (2002). *Accelerating Change: Rethinking Construction*. London: Strategic Forum for Construction.

Emmerson, HC (1962). *Emmerson Report; Survey of Problems Before the Construction Industries*. Ministry of Public Building and Works. London: HMSO.

Eriksson, PE and Lann, A (2007). Procurement effects on trust and control in client-contractor relationships. *Engineering, Construction and Architectural Management* 14(4): 387–399.

Farmer, M (2016), *Modernise or Die: The Farmer Review of the UK Construction Labour Market*. London: Construction Leadership Council.

Hawke, M (1994). Mythology and reality: the perpetuation of mistrust in the building industry. *Construction Papers of the Chartered Institute of Building*, 41(1994): 3–6.

HM Government (2010). *Low Carbon Construction Final Report (November 2010)*. London: HM Government. 52–62, 196–199.

HM Government (2013). *Construction 2025 – Industry Strategy: Government and Industry in Partnership*. London: HM Government. 23–25, 61–71.

HM Treasury (2010). *Infrastructure Cost Review: Main Report*. Infrastructure UK. Available at: www.gov.uk/government/uploads/system/uploads/attachment_data/file/192588/cost_review_main211210.pdf [Accessed 18th March 2015]

HM Treasury (2013). *National Infrastructure Plan*. Available at: www.gov.uk/government/uploads/system/uploads/attachment_data/file/263159/national_infrastructure_plan_2013.pdf [Accessed 18th March 2015].

Ikechukwu, U and Kapogiannis, G (2014). A conceptual model for improving construction supply chain performance. In Raiden, A B and Aboagye-Nimo, E (Eds), *Procs 30th Annual ARCOM Conference*, 1–3 September 2014, Portsmouth, UK: Association of Researchers in Construction Management. 1029–1038.

JCT (2014). *Setting the standard for construction contracts*. The Joint Contracts Tribunal. Available at: www.jctltd.co.uk/traditional-procurement.aspx [Accessed 1st December 2014].

Kaluarachi, DY and Jones, K (2007). Monitoring of a strategic partnering process: the amphion experience. *Construction Management and Economics*, 25(10): 1053–1061.

Latham, M (1994). *Constructing the Team*. London: The Stationery Office.

Lu, S and Yan, H (2007). A model for evaluating and applicability of partnering in construction. *International Journal of Project Management*, 25: 164–170.

MacKenzie, N and Tuckwood, B (2012). A model to manage the water industry supply chain effectively. *Management, Procurement and Law. Proceedings of the Institution of Civil Engineers*. Institution of Civil Engineers Publishing. 165(3): 181–192.

Maurer, I (2010). How to build trust in inter-organisational projects: the impact of project staffing and project rewards on the formation of trust, knowledge acquisition and product innovation. *International Journal of Project Management*, 28(2010): 629–637.

Morgan, S (2009). The right kind of bribe. *Building Magazine*, 9 October, pp. 8–9.

Potts, PG (1967). *Potts Report; Action on the Banwell Report: A Survey of the Implementation of the Recommendations of the Committee under the Chairmanship of Sir Harold Banwell on the Placing and Management of Contracts*. Economic Development Committee for Building of the National Economic Development Office. Available at: www.designingbuildings.co.uk/wiki/Construction_industry_reports [Accessed 15th December 2014].

Rahman, MM and Kumaraswamy, MM (2004). Contracting relationship trends and transitions. *Journal of Management in Engineering*, 20(4): 147–161.

RICS (2005). An exploration of partnering practice in the relationships between clients and main contractors. *Findings in Built and Rural Environments*. London: RICS Research. 2–3.

RICS (2012). *Contracts in Use. A Survey of Building Contracts in Use During 2010*. London: Royal Institution of Chartered Surveyors Publications.

Simon, E (1944). *Simon Report; the Placing and Management of Building Contracts*. Available at: www.designingbuildings.co.uk/wiki/Construction_industry_reports [Accessed 12th December 2014].

Tabassi, A, Ramli, M, Hassan A and Bakar, A (2011). Effects of training and motivation practices on teamwork improvement and improvement and task efficiency: the case of construction firms. *International Journal of Project Management*, 30: 213–224.

Vaaland, TI (2004). Improving project collaboration: start with the conflicts. *International Journal of Project Management*, 22(2004): 447–454.

Walker, A (2009). *Project Management in Construction*. 5th ed. Oxford: Blackwell Publishing Ltd. 150–158.

4 The importance of leadership for construction clients

4.1 Introduction

This chapter of the book essentially links academic theories around leadership theory with professional practice. Construction clients, in this regard, are seen as leaders on projects they procure. Accordingly, the purpose of forging this academic/professional practice link is intended to be a vehicle for improving leadership skills of construction clients. The chapter focuses on construction client leadership identity and highlights some aspects of leadership styles, broader practices and impact on individuals and organisations involved in construction projects. The review is primarily based on the leadership identity development (LID) model (Komives et al., 2009) to provide a critical review of leadership identity knowledge and understanding. In addition, it is based on its associated context within the framework of LID and is intended to improve outcomes for construction clients in leading and managing project teams.

Leadership definition, its transformative dynamics, the approaches to its applications, the challenges that come with it and its implications are critically reviewed using the LID model. According to this model developing leadership identity develops in three stages. The first stage highlights the awareness of the concept of leadership and the fact that some people lead and others follow. The second stage relates to the period of time that individuals gain experience and develop unique leadership styles first hand. The third stage relates to people learning the difference between management and leadership whilst at the same time experiencing leadership in practice.

4.2 Leadership identity and relevance for construction clients

> A leader is one who knows the way, goes the way, and shows the way.
> (Maxwell, 2006, as cited in Goodreads, 2018)

Arguably, social interactions of those individuals making up construction project teams require management, regulation and leadership to enable their creation, success and survival. Whilst project managers can provide this important intervention in the construction industry, the role of construction clients, as the creators and ultimately the owners or 'project sponsors', should not be underestimated. Temporary organisations, associated with development phases of projects, may be formed for the life of projects; others may already exist and passed on to future leaders to continue their development.

Leadership for construction clients should be focused on creating the right conditions and environment for people to succeed (Muller and Turner, 2010; APM, 2012; Francke, 2012). According to the LID model, leadership grows and matures through stages. Nurturing and supporting followers are an important and a distinctive feature of successful leaders. Lopes (2016) supported this argument and articulated that there are different ways to lead and motivate people, mostly by creating the right working environment and conditions. This has particular relevance for construction clients in selecting, motivating and developing project teams and providing the environment for them to procure successful construction outcomes.

There is wide acceptance that most 'lay construction clients', who perhaps have little or no previous construction management experience, knowledge or qualifications, learn from other professionals within project teams. Accordingly, it is not unusual for these relatively inexperienced construction clients to sometimes feel bewildered with the prospect of leading project teams. In such cases, they rely heavily on their construction consultants, especially at the early stages of projects. This may stem from the notion that less experienced individuals follow more knowledgeable people, who may be in positions of authority, power and legitimacy. In this regard, researchers such as Lord and Hall (2005) suggested that leadership in performance terms, through learning from others, progresses in stages from novice to intermediate and then to expert levels. They also claim that leadership skills are developed from micro levels through to problem and challenge based experience, and eventually to progressively a high level system influencing knowledge, attitude, behaviour and social perception. This 'learning pathway', influenced by lessons learnt from gaining experience, is sometimes referred to as an experiential model, gaining experience by doing using psychomotor domains, and is particularly relevant to how construction clients learn on projects.

It is important from the client's perspective to analyse some of the leadership skills that they need to develop in construction management. Leadership on projects can be regarded as a process whereby the interaction and activities of organised groups is planned and managed to achieve common goals (Buchanan and Huczynski, 2017). In the client context, it could be affiliated with Northouse (2015), who articulated that leadership is a complex process with multiple dimensions. Opinions on leadership have emerged from different disciplines, with many varying approaches to nurture the skills that are deemed necessary. Northouse also raised and challenged the subject of trait perspective from the viewpoint of 'born to be a leader', and argued that some people are born with characteristics and qualities to make them leaders. Leadership for construction clients could also relate to having the ability to influence groups of people to achieve prescribed goals (Newell and Grashina, 2004). This view is supported by Stogdill (1950, as cited in Buchanan and Huczynski, 2017) who defined leadership as a process to influence achieving common goals. In order to view this aspect from a psychological angle for construction clients, Knippenberg et al. (2005) emphasised that one of the principles of leadership is 'having influence'. Therefore, it is the impact and degree to which clients influence their construction teams, that leadership can be monitored and examined from a psychological perspective.

4.3 Leadership challenges for construction clients

Project management within the construction industry from clients' perspectives requires managing people, organisations, processes, systems, finance and the ever challenging goal of achieving success. In the leadership stakes, construction clients are often required to create the vision for a particular project, and then plan and execute it. In order to realise the dream and implement the vision, it can take many years through the planning stages to get all the stakeholders on board and get their 'buy in'. This can sometime present a daunting task for construction clients, especially those who have little or no prior experience of project procurement processes. During this important planning stage, construction clients frequently have to lobby politicians and government ministers to attract funding and get support and approvals through their respective local authorities. This is particularly the case for delivery of large scale masterplans, possibly linked to major regeneration, where funding is key to financial viability and affordability, and governance is paramount in ensuring best value for money is achieved. Programmes faced by construction clients can be challenging in this regard, against an environment of political uncertainly. Normally such large scale development involves a degree of cultural change within organisations, which presents another challenge for construction clients in gaining commitment and 'buy in' from staff within their organisations. Accordingly, change management processes are critical to the wellbeing of organisations and successful outcomes for projects. This notion is reinforced by Lopes (2016) who articulated that the key to successful leadership emanates from dealing with challenging situations, possibly related to staff resisting change, and gathering support for new initiatives. Clients therefore need to focus on any aspects related to change management, and consultation and discussions with key staff and therein dispel concerns and anxieties at early stages.

4.4 Key components of leadership for construction clients

In terms of leadership style, some would argue that construction clients need to be assertive, persuasive and articulate, but at the same time reasonable. For projects to succeed they should always be focused on 'win-win' outcomes for their respective client organisations and their contracting construction partners. Accordingly, those strategies, devised by construction clients, which create the right environment for collaboration amongst project teams, provide them with adequate and reasonable resources and get the stakeholders on board tend to achieve more successful outcomes. This argument seems to be in line with Fisher et al. (2017), which described such approaches to motivating others as 'identity based motivation theory'.

Walker (2007) argued that construction clients need to conduct themselves in a manner to enable them to get the best out of people who they manage, which is where leadership skills are crucial. This view is also endorsed by Northouse (2015), who suggested that leadership is similar to management due to commonalities between the two facets. He described a skills approach to leadership which shifts individuals' characteristics to skills learning and knowledge gain. This viewpoint enforces the perception that leaders are not born, but leadership can be taught to create leaders. Nonetheless, the debate centres on the question as to whether leaders who learn the skills and the 'tricks of the trade', but are without certain traits, can make good and exceptional leaders. Since research has informed that leadership and management are similar, questions are asked whether a good manager is good enough to be a leader? Chan and Drasgow (2001) claimed that non-cognitive credentials such as personality and values are based on individual differences that are linked to leadership performance indirectly. However, Miscenko et al. (2017) argued that an only skills based approach to leadership cannot cover all aspects of leader development, and this is particularly the case in construction clients. This is justified on the complex nature of leadership from a skills and identity basis, as articulated by Lord and Hall (2005).

4.5 Leadership styles as applied to construction clients

The Chartered Management Institute (CMI) defined leadership as 'the capacity to establish direction, influence and align others towards a common aim, motivate and commit others to action, and encourage them to feel responsible for their performance' (CMI, 2008, p. 1). It is also defined by Lester (2007, p. 303) as 'the ability to inspire, persuade or influence others to follow a course of actions or behaviour towards a defined goal'. Francke (2012) referred to the Centre for Creative Leadership's definition in that 'leadership is about creating the conditions for others to succeed'. This view has been supported by Muller and Turner (2010), who expressed the importance of creating a 'supportive environment'. This argument is also advocated by the Association for Project Management Body of Knowledge (APMBOK) in its quest for leadership to build the supportive environment for teamwork (APM, 2012). Lewis (2007) explained that there are hard and soft skills (human and technical elements), which need to be integrated with leadership and management skills, for achieving successful outcomes.

Successful leaders need to be good motivators and 'bring out the best in people' (Wong, 2007), and the importance of vision for leaders is paramount (Muller and Turner, 2010). Lewis (2007) believed that, as the majority of the managers' jobs often revolve around dealing with people, it is essential that people practice leadership alongside management. Furthermore, leaders hold power and authority over team members and knowing how to apply these is a major factor in maintaining morale (Ritz, 1994). Some would argue that the art of leadership is to make people to want to do things rather to be forced to do them. Milosevic *et al.* and Strohmeier (cited in Clarke, 2010) argued that the role of leaders should be focused on trying to get the best out of a spectrum of technical experts and professionals, which signifies leadership and effective management of team relationships and emotions. Opposing arguments have stated that exercising quality and competent leadership is the difference between leaders and dictators (Lewis, 2007). Along these lines, it might explain why leadership is a combination of personal characteristics and competency areas (Geoghegan and Dulewics, 2008). Furthermore, Abraham (2002) linked leadership to managing change within organisations and described it as a constant process to succeed, as leaders have to keep evaluating change and make improvements for future changes.

Arguably, the role of construction clients could, in the procurement of successful project outcomes, sometimes be too 'task focused' rather than 'people focused' which is a common misnomer when considering the aforementioned arguments around leadership. Most academic qualifications linked to construction management focus on the task nature of managing projects which are commonly referred to as 'hard skills'. Arguably, what construction clients mostly require, moreover, is people management skills often referred to by most commentators as 'soft skills'. Walker (2007) suggested that management is more of a mechanical process, whereas leadership is perceived to be forward thinking, charismatic and inspirational.

Buchanan and Huczynski (2017) articulated that there is a transitional period in leadership development identity. Furthermore, Galinsky and Kilduff (2013) explained a process whereby individuals become leaders or at least become regarded as leaders. This process takes into account that people go through a stage where there is mutual agreement of 'stepping up to the plate' and thus proving suitability to become a leader. Equally, this can be seen as an acceptance of the new leader by the group of people who are to be managed or the followers to be. Different perspectives exist, in relation to the distinction between management and leadership in this regard. Buchanan and Huczynski (2017) argued that 'managers do things right' and 'leaders do the right things', hence, leaders are seen to have a vision. This is particularly the case for construction projects and very important for construction clients to 'lead by example' to gain confidence from other members of the project team.

4.6 The importance of communication for construction clients in leadership

Academics have long argued that to become an effective leader, it requires not only to be a good communicator but also an outstanding one. Communications, in this regard, are defined as the art of transmission of ideas from an individual to another or a group of individuals with understanding (Newell and Grashina, 2004). Communication is key to understanding organisational behaviour (Buchanan and Huczynski, 2017), hence strong leaders need to be able to effectively convey their messages. This is particularly applicable to construction clients as leaders, in clearly articulating their organisational visions, as applied to project deliverables. They should also be able to persuade and influence their project teams to achieve common goals and purpose.

Creative leadership and the act of driving change in cultures, attitudes and behaviours in all aspects of business and/or organisations can be very challenging and frustrating for construction clients. However, if it is done well it can be very rewarding in driving transformation improvements. Knippenberg et al. (2005) debated that leadership is more effective when it is engendered in the leader's identity as well as a group identity as a collective character for the organisation. They suggested that consistent evidence exists to support that leadership is more effective when followers recognise collective identity and behaviour, and furthermore it suggests that followers as groups are more likely to endorse leaders who are group oriented.

Situational leaders can sometimes exist and emerge by a matter of choice, for example in the case of elected leaders in democratic societies. Construction clients who 'rise through the ranks' in their respective organisations could be an example of a similar path to leadership. Alternatively, at the other extreme, leaders can emerge by default or historical traditions such as monarchy systems or other means such as coercion, which unfortunately will lead to create autocratic leaders and dictators who are still sadly classed as leaders. In a similar scenario, construction clients can sometimes be those individuals who have inherited these roles as owners of the business where projects are being procured. It appears that the difference between different leaders may be the aspect of the use of power. Tourish (2013) presented a negative portrait of leadership as he argued that often leaders are too often regarded as heroes, charismatics and visionaries; and are held responsible for organisational success or failure. Whilst it can be argued that holding leaders accountable for success and/or the failure of any organisation is logical he articulated that leaders hold and use huge powers, which may not always be used fairly and wisely. There seems to be a fine line between getting the balance right to have a constructive directive leadership style, or an authoritative and autocratic style of leadership. Other balancing acts that one has to be mindful of are how much autonomy and participation is given to have a democratic or a laissez-faire style, and how this affects motivation.

It can be argued leadership amongst construction clients can address a wide range of management issues, and one of them is the process of change within their own organisations. The primary task for construction clients, as leaders, is managing and implementing change within their companies and dealing with the challenge of anticipating employees' reactions (Buchanan and Huczynski, 2017). Leadership for clients in the construction industry in its different forms and styles is about managing, directing, influencing and achieving common objectives and goals on projects. Leadership in this sense is a concept to lead and manage organisations to achieve change. Businesses and organisations need to keep changing, adapting and fulfilling corporate social responsibilities in the wake of globalisation and the ever increasing pressure on resources and commodities.

4.7 Adaptability of leadership styles for construction clients

Leadership skills are about motivating others, getting the best out of people (Wong, 2007). This notion is supported by Newell and Grashina (2004, p. 128), who concluded that leadership is a managerial interrelationship between leaders and followers and defined as the 'ability to influence groups of people in order to make them work and achieve prescribed goals'. Guillen et al. (2015) articulated that leadership drives organisation success and emphasised the importance of motivational measures and leading by example.

Construction clients are mostly senior individuals within organisations and normally accountable to the executive for the projects they become involved with, taking on additional responsibilities and getting involved with a wider range of people to manage directly or indirectly. One of the challenges this position brings is dealing with adversarial people from time to time and having difficult conversations with them, whilst managing workplace politics and organisational governance. This sometimes calls for construction clients to be courageous, diverse, dynamic and adaptive to different styles of leadership. Applications of leadership styles for construction clients can be complex and depend on the context of the situation they are under and the type of individuals they are dealing with. For this reason, there is no one size that fits all as far as approaches to the question of how to lead others. This view is attributed to the vast range of facets and complexity of leadership in leading and managing different type, structure, size, location and culture of different organisations. The findings of a study carried out by DeRue (2011, as cited by in Miscenko et al. 2017) suggested that leader development processes are complex and evolve over time. Chan and Drasgow (2001) reinforced this view in presenting a theoretical framework around individual differences and leadership. Their argument introduced a theoretical model around the integration of leadership performance processes and development. They assumed when leaders bring quality and personal characteristics to certain situations, this includes knowledge based attitudes and behaviours, learned and acquired alongside cognitive abilities and personality.

4.8 The relationship between poor client leadership and project failure

Lack of leadership displayed by construction clients can lead to many 'pitfalls' on projects. These can ultimately lead to project failures which can result in costly and disruptive implications for organisations. According to RICS (2016) the most common causes of project failures, from lack of leadership are as follows:

- Project initiation and planning

    - Failures to plan effectively 'if you fail to plan you plan to fail…'
    - Lack of clear project management and methodology

- Failures to understand project complexity and the effect on the probability of failure or success
- Lack of clear business objectives
- Poor briefing or changes to the brief
- Lack of understanding of project scope
- Lack of whole project life perspective

- Governance

  - Lack of ownership
  - Lack of leadership
  - Procedural issues
  - Avoidance

- People

  - Behavioural issues and attitudes
  - Breakdown of client and team relationships
  - Poor communications
  - Personnel changes and a lack of continuity throughout project
  - Lack of collaboration
  - Poor engagement with stakeholders
  - Bureaucracy
  - Political aspects and barriers
  - Inexperienced project sponsor/construction client

- Technical

  - Lack of resources
  - Poor brief
  - Failure to understand risks
  - Procurement and finance difficulties

Throughout the book, assistance for readers will be provided to addressing most of the aspects above, to improve their leadership skills. This is a concerted attempt, by the authors, to support construction clients and avoid failure of projects wherever possible. Measures such as construction clients adopting the right project controls, governance, communication strategies, and collaboration in place and being competent leaders will be discussed in later chapters.

4.9 Summary

This chapter has referenced current academic literature on theories relating to leadership and applied these to the roles of construction clients. As mostly leaders on projects, it is essential for these individuals to understand what makes a good leader, in order to perform their roles and motivate others for achieving successful outcomes. Accordingly, in those positions of leadership, construction clients require courage, conviction and assertiveness when faced with difficult choices and decisions. Their decision making is subject to ethics, fairness and principles which require to be considered and reflected in the decision making process. The actions of construction clients in this regard, can have consequences for the vested interests of their employees, communities and the wider society and decisions made can have far reaching consequences and devastating effects. Leadership is about how to 'know'; first of all to know themselves, their competitors, friends, enemies, available resources and their own capabilities. Once they know the way, they can lead on the construction process to benefit the build outcomes, the design team and the project environment. Construction clients attribute this to the realisation that being a leader is far more challenging than originally envisaged.

The leadership identity development and progression process as applied to construction clients, is dependent and influenced by individual differences, cognitive capacity, personality and temperament, personal identity, personal values and emotional intelligence, driven by cultural context and personal experience.

Finally, there are many areas where lack of leadership displayed by construction clients can lead to failures on projects. These have been categorised as poor project planning, lack of clear governance, behavioural issues and communication/collaboration difficulties.

References

Abraham, GL (2002). *Identification of Critical Success Factors for Construction Organization in the Architectural/Engineering/Construction (AEC) Industry*. Georgia: ProQuest.
APM (2012). *Project Management*. Available at: www.apm.org.uk [Accessed 19th January 2017].
Buchanan, DA and Huczynski, AA (2017). *Organisational Behaviour*. 9th ed. London: Pearson.
Chan, K-Y and Drasgow, F (2001). Toward a theory of individual differences and leadership: understanding the motivation to lead. *Journal of Applied Psychology*, 86, 481–498.
Clarke, N (2010). The impact of a training programme designed to target the emotional intelligence abilities of project managers. *International Journal of Project Management*, 28, 461–468.
CMI (2008). *Managing Projects*. Corby: Chartered Management Institute: Checklist 035.
Fisher, O, O'Donnell, SC and Oyserman, D (2017). Social class and identity-based motivation. *Current Opinion in Psychology*, 18, 61–66.
Francke, A (2012). Briefing. *Professional Manager*, p. 5.
Galinsky, A and Kilduff, G (2013). Be seen as a leader. *Harvard Business Review*, 91(12), 127–130.
Geoghegan, L and Dulewics, V (2008). Do project managers' leadership competencies contribute to project success?. *Project Management Journal*, 39(4), 58–67.
Goodreads (2018). *Goodreads*. Available at: www.goodreads.com/quotes/ [Accessed 25th January 2018].

Guillen, L, Mayo, M and Korotov, K (2015). Is leadership a part of me? A leader identity approach to understanding the motivation to lead. *The Leadership Quarterly*, 26, 802–820.

Knippenberg, B v, Knippenberg, D v, Cremer, DD and Hogg, MA (2005). Research in leadership, self, and identity: a sample of the present and a glimpse of the future. *The Leadership Quarterly*, 16, 495–499.

Komives, SR *et al.* (2009). Leadership identity development: challenges in applying a developmental model. *Journal of Leadership Education*, 8(1), 11–47.

Lester, A (2007). *Project Management Planning and Control*. 5th ed. Oxford: Butterworth-Heinemann.

Lewis, JP (2007). *Fundamentals of Project Management*. 3rd ed. New York: AMACOM, American Management Association.

Lopes, PN (2016). Emotional intelligence in organisations: bridging research and practice. *Emotion Review*, 8, 1–6.

Lord, RG and Hall, RJ (2005). Identity, deep structure and the development of leadership skill. *The Leadership Quarterly*, 16, 591–615.

Miscenko, D, Guenter, H and Day, DV (2017). Am I a leader? Examining leader identity development over time. *The Leadership Quarterly*, 28, 605–620.

Muller, R and Turner, R (2010). Leadership competency profiles of successful project managers. *International Journal of Project Management*, 28(5), 437–448.

Newell, MW and Grashina, MN (2004). *The Project Management Question and Answer Book*. New York: AMACOM, American Management Association.

Northouse, PG (2015). *Introduction to Leadership*. 3rd ed. Los Angeles, CA: Sage.

RICS (2016). *Lessons Learned*. Available at: www.rics.org [Accessed 23rd June 2018].

Ritz, GJ (1994). *Total Construction Project Management*. Boston, MA: McGraw-Hill.

Tourish, D (2013). *The Dark Side of Transformational Leadership: A Critical Perspective*. East Sussex: Routledge.

Walker, A (2007). *Project Management in Construction*. Oxford: Blackwell.

Wong, Z (2007). *Human Factors in Project Management*. San Francisco: Jossey-Bass.

5 Governance considerations for construction clients

5.1 Introduction

This chapter predominantly is intended to make construction clients aware of some of the governance requirements in respect of their projects. It will explain some of the processes around project approval and the importance of project/programme boards in this regard. Various examples of templates and processes for various stages of the approval process will be given to assist construction clients understand the level of information required.

5.2 Project controls

For construction clients, it is imperative that they understand their roles, responsibilities and authorisation levels when leading their project teams. This will ensure they do not make decisions which have a financial or operational effect on projects which should have been escalated to a higher authority. Normally authorisation levels are established for a given project, or they mirror financial regulations for that client body. For construction clients, having robust decision making processes whilst ensuring governance procedures are maintained is fundamental; avoiding unnecessary delays whilst not infringing financial regulatory requirements.

5.3 The importance of project/programme boards

The main purpose of these boards is to provide governance and transparency to projects, and avoiding the responsibility and accountability for decision making falling to one individual. Having project and programme boards should facilitate a cross disciplinary approach to managing the governance of projects, which is especially important when large capital investment is involved. Furthermore, it also allows for additional resources to be committed to and creates joint accountability for decision making and project progress.

It is normal for project boards to approve various stages of the project and any major variations and deviations. Accordingly, decisions can be taken by those individuals that make up the board depending on the financial impact. In some cases, if the value of a decision making process exceeds the authorisation level of the project board then the decision making needs to be referred to a higher level. Under normal situations, this higher level could be the client executive team. Notwithstanding this, some public sector bodies, which can include universities or local authorities, sometimes have a hierarchy of boards that need to be reported to for seeking project or programme approval.

One example, in the context of the university sector, is illustrated in Figure 5.1 which shows a Project Board reporting to a Programme Board, which in turn reports to a Finance and Resources Committee and ultimately a Governing Council/Board of Governors.

Limits of authority for each board will be clearly set down in governance and financial regulations. An example for levels of authority is contained in Table 5.1.

Construction clients need to be mindful when their projects require a higher tier of approval within their respective organisations, that they allow sufficient time in their project programme for the approval process. The reason for this, is that normally higher level boards will only consider those capital projects that have been approved by lower boards, and for which robust business cases and reports recommending reasons for approval are clearly set down. Where some boards are not convened on a frequent basis, e.g. every quarter, project planning around critical milestones where approval needs to be received is crucial for avoiding programming delays.

5.4 Gateway processes for project approval and business cases

In considering the approval processes for projects through the individual boards, it is commonplace for robust and financially rigorous business cases to be prepared to support the business venture. In some organisations, mostly public sector, there may be 'gateway processes' which require staged approval depending on the stage and development of projects. An example, taken from a UK university governance process, could be as follows:

- Obtain Gateway 1 approval. Normally this would be at the RIBA Stage 1, once an initial feasibility study has been prepared and construction clients are seeking 'in principle' approval to proceed to next design stages, based on feasibility costs, programme and outputs. At this early stage, predictability in terms of design intent, costs and programme may be indicative, and subject to further design development. This is normally supported by an outline business case setting out the various options and the benefits and possibly implications of the preferred option that the gateway seeks approval for. An example of a template for the Gateway 1 process is contained in Appendix 1A.
- Obtain Gateway 2 approval. Normally this would be at the end of RIBA Stage 2, once initial feasibility studies have been developed into more

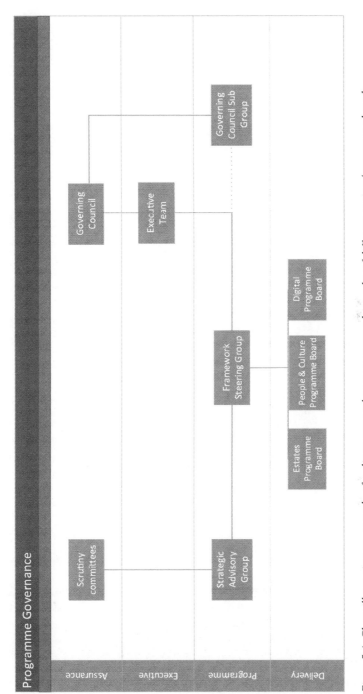

Figure 5.1 Chart illustrating an example of a client approval process across a hierarchy of different project/programme boards

Table 5.1 Example of financial levels of authority for a client organisation

Proposed item for approval	Financial limits
Acquisition, construction, refurbishment and disposal of buildings (and acquisition and disposal of land) and equipment, included in three-year outline capital budget approved by Governing Council	◊ Governing Council >£5m ◊ Executive £1m–<£5m ◊ Programme Board £100k–<£1m ◊ Finance Director/Director of Estates & Facilities <£100k
Acquisition, construction, refurbishment and disposal of buildings (and acquisition and disposal of land) and equipment, in **addition** to that in three-year outline capital budget approved by Governing Council	◊ Governing Council >£1m ◊ Executive £100k–<£1m ◊ Programme Board – for info ◊ Finance Director/Director of Estates & Facilities <£100k
Acquisition, construction, refurbishment and disposal of buildings (and acquisition and disposal of land) and equipment, by **substitution** to that in three-year outline capital budget approved by Governing Council	◊ Governing Council >£1m ◊ Executive – for info ◊ Programme Board £100k–<£1m ◊ Finance Director/Director of Estates & Facilities <£100k
Leasing of properties	◊ Governing Council >£5m ◊ Executive £1m–<£5m ◊ Programme Board £100k–<£1m ◊ Finance Director/Director of Estates & Facilities <£100k
Approval of building contracts (Subject to approved business case and undertaken prior to raising of Purchase Order)	◊ Finance Director/Director of Estates & Facilities >£5m ◊ Associate Director of Estates & Facilities <£5m (and in all scenarios sign-off from Legal Services)
Approval of increase in approved budget for capital projects	◊ Up to the lesser of 10 per cent or £100k to be approved by the Finance Director/Director of Estates & Facilities

detailed plans and designs. It would normally be expected that 'composite cost' plans have been prepared at this stage which have estimated costs on a square metre rate, to give approximately 60 per cent cost certainty. It is possible at this stage that more developed 'elemental costs' have been prepared which are based on more detailed cost breakdowns (by elements) and give approximately 70 per cent cost certainty. The extent to which the costs have to be developed for a Gateway 2, really depends on the governance and financial policies set down by a specific organisation. At this stage, construction clients are seeking approval to proceed to full detailed design stages, which normally implies a commitment to approve the remainder of

the pre-construction professional design team fees. This is normally supported by a more developed business case.
- Obtain Gateway 3 approval. Normally this would be at the end of RIBA Stage 4, once the detailed design has been completed and tender prices have been obtained. It would normally be expected that tender costs are fixed and based on detailed cost and tender documents for to give 100 per cent cost certainty. At this stage, construction clients are seeking approval to proceed to the construction phase and, as such, a commitment to approve the whole project costs. Normally the 'whole project costs' include all associated costs including VAT, professional fees, decant costs, contingencies, equipment and fixtures and fittings. This Gateway 3 as the final gateway must provide a full business case, setting out a developed and robust business case with financial rigour and strong arguments for the development to proceed, having considered all other available options. An example of a template for the Gateway 3 process is contained in Appendix 1B.

When considering the above Gateways 1, 2 and 3, the flowchart in Figure 5.2 could be helpful in illustrating how each of the gateways is integrated into an overall process for approving projects, at the various board levels and design stages.

5.5 Summary

This chapter has articulated the importance of construction clients having robust financial processes in place. Accordingly, it is imperative that they understand their roles, responsibilities and authorisation levels when leading their project team. Furthermore, it is important for construction clients to have robust decision making processes in place within their respective organisations whilst ensuring governance procedures are maintained.

Project and programme boards are normally the vehicle by which the decision making process is governed. Their main purpose is to provide governance and transparency to projects, and avoiding the responsibility and accountability for decision making falling to one individual. With this in mind, where the value of a decision making process exceeds the authorisation level of the project board then the decision making needs to be referred to a higher level. Accordingly, construction clients need to be mindful when their projects require a higher tier of approval within their respective organisations, that they allow sufficient time in their project programme for the approval process.

In considering the approval processes for projects through the individual boards, it is commonplace for robust and financially rigorous business cases to be prepared to support the business venture. In some organisations, mostly public sector, there is are gateway processes for project approval and business cases. The gateway processes will highlight different levels of approval for projects depending on which design stage they are at. Practical examples and templates of gateway applications, supported by business cases, have been presented in this chapter, to assist readers in this regard.

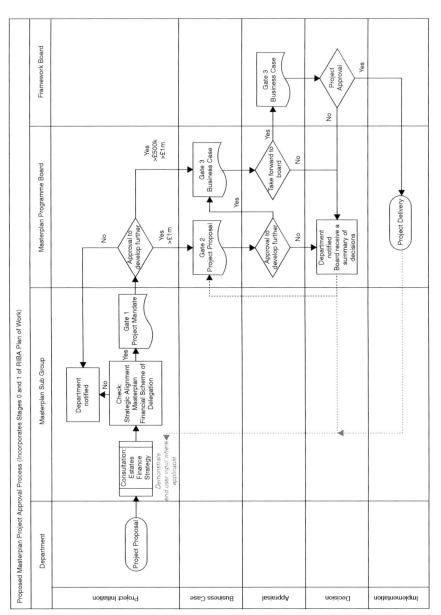

Figure 5.2 Flowchart to illustrate a gateway approval process

6 Selection and appointment processes for construction clients

6.1 Introduction

This chapter highlights the importance for construction clients of the processes around selection and appointment of their contracting partners. Having the right main contractors on board is arguably one of the most important aspects for construction clients. Accordingly, this chapter articulates some of the issues that clients need to be aware of when commencing their selection processes. This is intended as a means of bringing rigour to existing processes and ensuring that a staged approach to the appointment of project teams is not compromised. Furthermore, it articulates the perceived problems that construction clients face in selecting the right contracting partners, especially in the case of two stage tender procurement. It also provides a historical perspective on this overarching dilemma for construction clients, with particular reference to government reports.

This chapter is focused predominantly on a research study by Taylor (2017), which established what information and methods are available to construction clients to help them to identify and select suitable contracting partners. This was aimed at promoting the choice of the right partners for a particular project and encouraging long term collaborative relationships for achieving successful project outcomes. Previous opinions had suggested that construction clients may not be employing the full use of robust pre-qualification methods in their appointment processes. This could possibly be hindering their chances of selecting the most suitable partners which could ultimately be adversely affecting project outcomes. In this regard, the study provided important findings to support the following areas of research:

- Determination of whether the quality of construction clients' pre-qualification processes influence project success.
- Determination and analysis of whether the current pre-qualification practices of construction clients are sufficiently robust to identify and select the right contracting partners.
- Identification of the extent to which construction clients are using objective assessments of key performance indicators (KPIs) to shortlist contractors for tenders.

6.2 The importance of the contractor selection process

Critchlow (1998, p. 24) stressed the importance of the selection processes when using collaborative procurement strategies, in choosing the most appropriate contractors to realise benefits of partnering through pro-activity, building team spirit, employing lateral thinking and exploring alternatives. Gulf Construction (2008, p. 1) reinforced this view and advocated that the selection processes should carefully evaluate contractors' experience, skills, resources and expertise rather than simply appointing contractors on lowest tender price. Furthermore, Governments' Procurement Group advocated that selection criteria and competition should be on the basis of life cycle value for money rather than simply reliant on initial capital costs (Procurement Group, 1999, p. 1). This view was supported by Kwakye (1991), who advocated that owing to contractors' early involvement with project teams, selection should be firmly based on value rather than lowest tender. Egan also underlined this opinion and stated that: 'too many clients are indiscriminating and still equate price with cost, selecting designers and contractors exclusively on the basis of the tendered price' (Egan, 1998, p. 6).

Selection criteria was also referred to in the Code of Procedure for Two Stage Tendering which recommended, at the short listing stage, that contractors should be evaluated upon financial stability, competence, general expertise and experience, approach to quality assurance, capacity and the company's technical and management structure (National Joint Consultative Committee for Building, 1996, p. 5). In addition, the UK Governments' Procurement Group suggested that incentives should be incorporated into building contracts to promote the extent and degree of collaboration in yielding further benefits and improvements beyond those included for within the agreement itself (Procurement Group, 1999, p. 1).

Farmer (2016), highlighted that there was still a 'collaboration problem' in the UK and its adversarial approach remains a salient issue, which makes the industry resistant to change. Despite a succession of government and academic reports, Latham (1994) *et al.*, anecdotal evidence would appear to indicate that whilst the majority of private clients have accepted the need for, and the potential advantages of partnering and early contractor involvement, many are struggling to find trustworthy partners. This raises questions as to whether clients are utilising suitable pre-qualification selection criteria. Turskis (2008) has suggested that many clients undervalue the importance of robust pre-qualification methods or, in some instances, overlook them altogether. Huang and Wilkinson (2013) stressed the importance of what they refer to as 'prior learning' as a trust building tool, especially for new relationships. Good quality pre-qualification processes increase the likelihood of successful project outcomes, however, many prominent researchers in the field of pre-qualification (Holt *et al.*, 1995) generally concurred that many practitioners carry out only subjective pre-qualification checks and some do not carry out any kind of formal review of contractors at all. Ng *et al.* (2002) explored and proposed a concept that would allow contractors' key performance indicator (KPI) data to be collected, verified and disseminated to clients via a web based platform.

The study by Taylor (2017) carried out a questionnaire survey of 31 construction procurers, with the objective to assess common pre-qualification practices and the subsequent effects on project outcomes. Statistical analysis and tests were carried out which confirmed that robust pre-qualification practices do improve the chances of achieving successful project outcomes. However, the findings also showed that the full range of good quality pre-qualification processes available to practitioners are not used consistently. Other findings included: practitioners frequently fail to follow up with contractor reference checks, do not always use objective assessments of contractor key performance indicators (KPIs), and the use of BIM improves collaboration.

6.3 Articulation of the problem for selecting contracting partners from the perspective of construction clients

There is a perception amongst private clients and procurers of large construction projects that the majority of contractors are still not truly embracing a culture of open and collaborative working. Furthermore, many clients believe that some opportunistic main contractors are using early involvement in projects as a means to force unfairly inflated prices and risk onto clients, effectively holding clients to ransom in the knowledge that they do not have the luxury of time to re-tender the works. Despite a succession of government and academic reports, anecdotal evidence would appear to indicate that, whilst the majority of private clients have accepted the need for, and the potential advantages of partnering and early contractor involvement, many are struggling to find trustworthy partners whom they are willing to give repeat contracts to. This raises questions as to whether clients are utilising suitable pre-qualification selection criteria. Turskis (2008) suggested that many construction clients undervalue the importance of robust pre-qualification methods or in some instances overlook them altogether. The effects of clients selecting the wrong contracting partners are that schemes can suffer from viability issues leading to large delays in the pre-construction phase or in some instances with projects being jettisoned. Some construction clients have either incurred long implementation delays or have introduced a viability section in their planning processes. Some projects have come to fruition but concluded with the perceived non-delivery of the project objectives by either contracting party. Others have resulted in conflicts and disputes which represents a wasteful use of resources and an unnecessary burden to the industry's productivity problems (Latham, 1994). If there are widely held perceptions that projects that have been negotiated and have had early involvement from main contractors are suffering from unfairly inflated prices, why is it that contractors are not all making high profit margins? In the past year, 'we have had a litany of poor results from major supply-chain players' (Farmer, 2017), which culminated in Carillion, the second largest main contractor in the UK, going into liquidation after amassing huge debts of around £1.5 billion in 2018 (BBC, 2018). Could it therefore be construed that if construction clients are indeed paying inflated prices and contractors' profit margins, there is a compelling problem with the performance of construction projects?

6.4 A historical perspective of the problem

Tendering originated from pre-contract communication between architects and builders. By the end of the eighteenth century, architects' roles were consolidated: into construction designers and 'leaders' of project coalitions, hence establishing 'traditional procurement' (Holt et al., 1995). Traditional procurement generally refers to procurement routes that involve the splitting of the design phases from the construction phases. Design teams in this scenario are employed directly by construction clients, to both design and administer construction contracts. Normally, upon completion of the design processes and after competitive tender, contractors are employed to execute the construction works. Hence, with the exception of the quality of the workmanship and adherence to the drawings and specifications, contractors bear no responsibility for the suitability of the designs. Historically this method is widely recognised as being adversarial in nature and prone to contractor claims. Furthermore, as contractors are effectively excluded from the design processes, projects can suffer from buildability issues, which can have a detrimental effect on time, quality and cost. A Government committee report by Simon (1944, cited in Holt et al., 1995) examined and recommended suitable procurement methods to facilitate urgently needed new social housing stock and public buildings such as schools and civic buildings on a scale never previously, or for that matter since, seen. Simon made some far reaching observations and recommendations around selective tendering. He articulated if companies are shortlisted, based on the quality of the service or goods that they provide, rather than open tendering based on lowest price, it was much more likely to ensure that the products fulfil clients' needs, for a fair price. It is generally accepted that the Simon Committee initiated the move away from open tendering in lieu of encouraging negotiated procurement; therein allowing contractors to become involved with the project at the design stages (Holt et al., 1995). It is, however, widely accepted that the recommendations made by Simon were not implemented by either the incumbent or successive governments during the next 20 years (Costain, 1964). In 1964 the government commissioned a further report when Sir Harold Emerson was asked by the Minister of Public Building and Works to consider practices adopted for the placing and management of construction work, and to make recommendations for promoting efficiency and economy within the industry (Holt et al., 1995). The report led to the formation of a committee led by Banwell (1964) after who the report became to be known by. The recommendations put forward by Banwell echoed many of those made by Simon and whilst unlike the apathy that followed the publication of the Simon Report, some of the recommendations were implemented by the government; they were not wholly incorporated into government procurement methods and were largely ignored by the private sector. It was not until the 1990s following the release of the two seminal government reports by Latham (1994) and Egan (1998) (both of whom were calling for root and branch transformations) that the industry as a whole began to accept that there was a need for change towards collaborative procurement and working methods. However, some 11 years after Egan made his recommendations, Wolstenholme (2009) who was commissioned by Constructing Excellence to review progress within the sector, found that whilst there had been some progress, many attempts to change had been superficial and most of Egan's targets had not been achieved (EC Harris LLP, 2013). The Farmer Report (Farmer, 2016) that was commissioned by the government to review the construction labour model vis-à-vis skills shortages highlighted that there was still a 'collaboration problem' in the UK and its adversarial approach remains a salient issue which makes the industry resistant to change.

6.5 Risk

> No construction project is risk free. Risk can be managed, minimized, shared, transferred or accepted. It cannot be ignored.
>
> (Latham, 1994, p. 14)

Whilst the subject of risk is to a large extent beyond the scope of this chapter, it is important for construction clients to understand the dynamics and approach to risk management in construction procurement, as this can be a major factor in engendering collaborative professional relationships (McDonald, 2013). Furthermore, 'risk' is an inherent factor of all construction projects, as they are often unique, constrained and are subject to the vagaries of human characteristics and shortcomings (APM, 2006, p. 26). One of the major difficulties with risk management for construction clients is deciding and importantly, agreeing, which party will be responsible for managing specific risks and, perhaps more pertinently, bear the consequences if the risk manifests. The consensus is that the risks should be allocated to the party best able to control them and should be balanced proportionally to the extent of the stakeholder's interests in the project (Lehtiranta, 2014, p. 642). Unfortunately opportunism can be prevalent in construction projects as parties often have different organisational objectives and targets. Traditionally opportunistic behaviour has been countered by utilising control oriented mechanisms but this can be counter-productive to engendering collaborative relationships (Osipova and Eriksson, 2013, pp. 391–392). This raises the importance of trust between the parties: 'Absence of trust and collaboration in risk management led to a low level of risk communication during the procurement phase' (Osipova and Eriksson, 2011, p. 1154).

An interview of UK main contractors for a study by Laryea and Hughes (2008) found that most of the respondents were not using formal analytical risk probability models such as Monte Carlo simulation, but rather relied on intuition and experience to identify risks and quantify the likelihood of occurrence. When asked about common causes of increased risk allowances, the respondents mentioned short tender durations, incomplete designs, buildability issues and clarity of tender documents. As contractors are generally employed early in two stage tendering procurement methods and are tasked with managing or at the very least participating in the design development, these issues should be less prevalent (Naoum and Egbu, 2016) hence this could indicate that the comments relate to other types of tendering arrangements.

The government's drive for increased use of Building Information Modelling (BIM) should also reduce issues that arise from buildability problems (Naoum and Egbu, 2016, p. 325) and according to *Constructing Excellence* (BRE, 2018) this is one of the major practices that engenders collaborative working. However, Vass and Gustavsson (2017) opined that the industry and particularly private clients have been slow to adopt the use of BIM. Research figures listed by Ghaffarianhoseini *et al.* (2017, p. 1049) indicated that participation rates for use of BIM in UK construction in 2017 were 39 per cent, and asserted that the public sector accounts for approximately 40 per cent of construction output at this time. As the government has been a major proponent of BIM since 2011, and made its use mandatory on all public funded works in 2016 (Lindkvist, 2015), these statistics would appear to suggest that use of BIM in the private sector is low.

6.6 Benchmarking

> We do not need to measure everything that matters; we only need to measure the things that matter.
> (Saad, 2005, as cited in Raymond, 2008)

For construction clients to be able to undertake the selection processes for their consultants and main contractors, it is important for them to be able to benchmark each contender against certain criteria. This enables them to measure using such criteria and weighting each submission, based on what they believe is important for them and their projects.

According to McCabe (2001, p. 27), the following definition by McGeorge and Palmer (1997) provided the most succinct but complete description of the essential features of benchmarking:

> A process of continuous improvement based on the comparison of an organisation's processes or products with those identified as best practice. The best practice comparison is used as a means of establishing achievable goals aimed at obtaining organisational superiority.
> (McGeorge and Palmer, 1997, as cited in McCabe, 2001, p. 27)

Benchmarking, however, has many different variations and uses as a management concept, depending on whether companies are comparing themselves against their peers or just comparing competing contractors as part of a selection process (Cowper and Samuels, 1996). With regard to the selection of suitable contractors: 'the comparison between one company and another may depend on performance benchmarking' (Luu *et al.*, 2008, p. 759). Luu *et al.* (2008 p. 760) stated that benchmarking is a comparison of the performance of external organisations to try to improve the organisation's own performance. Rigby *et al.* (2014) described benchmarking as an 'information-based approach to performance management'. Saad and Patel (2006) suggested that benchmarking is a process of 'measuring, comparing, learning and improving'.

According to *Constructing Excellence* and BRE (2018), companies carrying out benchmarking exercises should avoid trying to measure too much initially and should first focus on the key salient performance issues.

Latham (1994) suggested that a national database be established (Sutton, 2008), and Egan (1998) concurred: recommending benchmarking as a tool, both to help contractors improve their own performances and also as an approach to assist clients with contractor selection.

Benchmarks can also be used to allow procurement organisations or clients to assess the performance of construction firms. Here information can be accumulated either by clients solely, based on their experiences or it can be requested from suppliers to help clients draw a detailed picture of their suppliers' performance (Rigby et al., 2014, p. 787).

Constructing Excellence (BRE 2018) suggested that there are five key steps in the benchmarking process:

> **Plan:** Clearly establish what needs to be improved – make sure it is important to you and your customers – and determine the data collection methodology to be used including any key performance indicators (KPIs).
> **Analysis:** Gather the data and determine the current performance gap – against a competitor, the industry or internally – and identify the reasons for the difference.
> **Action:** Develop and implement improvement plans and performance targets.
> **Review:** Monitor performance against the performance targets.
> **Repeat:** Repeat the whole process – benchmarking needs to become a habit if you are serious about improving your performance.

6.6.1 Key performance indicators

> A Key Performance Indicator (KPI) is the measure of performance of an activity that is critical to the success of an organisation.
>
> (BRE, 2018)

The precise history of performance measurement is not known but the practice was used by the emperors of the Wei Dynasty in the third century AD to benchmark its family members. It is thought that its first use in industry was possibly by Robert Owen in the early 1800s in the Scottish mill industry (Banner and Cooke, 1984, p. 328). Dr W. Edward Deming who was credited as being instrumental in the post-war economic recovery in Japan was a prominent advocate of performance measurement. He was also a firm believer that focusing on the customer was the key to improving performance (McCabe, 2001, p. 27).

Benchmarking is identifying 'best practice' used by others and comparing and contrasting the results of organisations using a set of predetermined key performance indicators (Fernie et al., 2006). Latham (1994) highlighted the benefits of benchmarking and suggested that a national forum be established and an ombudsman appointed to monitor the performance of the industry. Many of Latham's recommendations were reiterated by Egan (1998) but interestingly Egan also suggested that a score card system be adopted and the names of the best performers made public. A national KPI working group was established which set out a framework by which the industry could measure its performance (The KPI Working Group, 2000):

> Sir John Egan's report, Rethinking Construction, challenged the industry to measure its performance over a range of its activities and to meet a set of ambitious improvement targets. This is the KPI Working Group's answer to that challenge. It sets out a comprehensive framework which construction enterprises can use to measure their performance against the rest of the industry, and has been designed to be used by organisations, large or small, specialist or supplier, designers or constructors ... The KPI Pack presents the construction industry's range of performance by presenting organisations with a framework to benchmark activities both at a broad level, and at a level much closer to the 'coal face' – such as rectifying defects and meeting clients' expectations.
>
> (The KPI Working Group, 2000)

The BRE, through its 'Constructing Excellence' programme and in conjunction with Glenigan, collected data from the construction industry and allowed subscribing members to monitor their own performance against the industry average (BRE, 2018). Glenigan had published annual construction performance reports, however in respect to performance criteria, the data is summarised and anonymous therefore reflected only the results of the industry averages. It appeared to have little to offer to the enquiring procurer but is aimed at providing contractors with industry average performance data from which to benchmark themselves against. However, studies by Ng et al. (2002), Beatham et al. (2004) and Rawlinson and Farrell (2010) suggested that the industry as a whole has also under-utilised the use of KPIs as an improvement tool. In response to the Latham and Egan Reports, the UK construction industry has developed its own set of KPIs. However, their effective use has been limited (Beatham et al., 2004, p. 93).

6.6.2 Constructionline

Constructionline was set up by the government in 1998, initially as an approved contractor database for the Ministry of Defence contracts (Steele et al., 2003). This was in response to the recommendation made by Latham

(1994), who articulated that duplicating pre-qualification practices was a wasteful practice (Beales, 2008). The government put Constructionline up for sale in 2014 (Mann, 2014), and it was purchased by Capita in January 2015 for £35 million (Dunton, 2015). Capita had been operating the platform for a number of years under a concession agreement (Mann, 2014). It seemed that whilst Constructionline has been widely used for public contracts, the private sector has not wholly embraced the platform, as is evident by a group of main contractors only agreeing to adopt its use in 2014 for subcontractor pre-qualification health and safety criteria some 16 years after it was established (Construction News, 2014). Bouygues UK (2018) stated that 'We're proud to be part of an initiative that seeks to save suppliers in the industry more than £25 million a year.'

It is apparent that Constructionline reduced bureaucratic red tape and unnecessary costs incurred in the duplication of submitting documentary information, particularly for many small and medium enterprises (SMEs) when responding to pre-qualification questionnaires (PQQs) as part of the bidding process (Leflty, 2014). It was, however, evident that it did not become the score card league table that Latham (1994) envisaged.

The application process uses the Publicly Available Specification 91 (PAS 91) questionnaire that was established by the British Standards Institute (BSI) in response to the government's desire to streamline construction procurement PQQs (Constructionline, 2018). PAS 91 is mandated by the government, however it is not classified as a British Standard. Reeve (2013) suggested that, whilst it would require lengthy consultations, if it was given full British Standard status and possibly even made into a specification, it would be more likely to be fully adopted by the industry as a whole.

6.6.3 *Centralised KPI sharing systems*

Ng *et al.* (2002) explored and proposed a concept that would allow contractors' KPI data to be collected, verified and disseminated to clients via a web based platform. They argued that such a system would allow registered construction clients to link individual contractors' performance results to its own goals and objectives and streamline the process of contractor selection.

The information would be collected from construction clients (the appraisers) whom would rate contractors' key performance criteria as highlighted in Figure 6.1. It is acknowledged that there could be potential difficulties in obtaining accurate and fair appraisals. To overcome such challenges they proposed that before scores and rankings were published the data would be vetted and checked by an independent panel and that contractors would be given the opportunity to appeal the results. Ng *et al.* (2002) cited studies by Palaneeswaran and Kumaraswamy (2001) as having demonstrated how a similar system of Contractor Performance Assessment Reporting System (CPARS) had been implemented and had proved successful by government clients in the US. However, Bradshaw and Chang

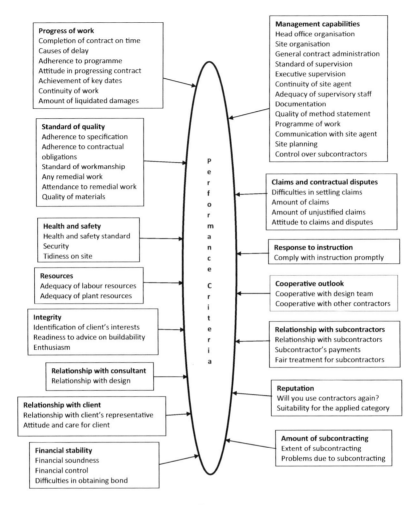

Figure 6.1 Performance criteria and subcriteria

(2013) asserted that the marking criteria of CPARS can be inconsistent. This was supported in the following quotation:

> Observations show that program managers and contracting officers are often reluctant to report negatively on past performance because this can reflect poorly on their own ability to manage the program or contract. Furthermore, to avoid conflict with the contractor, the government may refrain from documenting performance deficiencies in official databases. As a result, the past performance write-up does not always reflect a contractor's performance accurately.
>
> (Bradshaw and Chang, 2013, p. 75)

It is somewhat unsurprising that contractors in the USA have been known to litigate to contest appraisals by government contracting officers (Reed, 2018), so this perhaps highlights the challenges of implementing a similar system for the private sector in the UK.

6.6.4 Common pre-qualification practices

> An effective selection process is crucial for clients wishing to strike a balance for successful project outcomes.
>
> (Fong and Choi, 2000)

The Government's *Construction 2025* report (HM Government, 2013) reinforced this view and advocates that selection processes should carefully evaluate contractors' experience, skills, resources and expertise rather than simply appointing contractors on lowest tender (Challender et al., 2014, p. 1041).

One of the most important activities for construction clients is the selection of main contractors; and the use of robust pre-tender or pre-qualification selection methods which influence the likelihood of achieving clients' objectives and successful project outcomes (Hatush and Skitmore, 1997b). Peace (2008, p. 33) supported this view adding 'the selection and appointment of construction firms is one of the most important steps the client's internal team take to ensure a project's success'. Peace also stated that data collected during the evaluation phase should be assessed systemically and decisions taken on an objective basis. However, Doloi (2009, p. 1247) indicated that in the UK there was an overreliance of subjective analysis of whether contractors would likely be able to successfully deliver projects and meet clients' objectives. This view was supported by Turskis (2008, p. 225) who argued that the selection of contractors is often based on intuition. Accordingly, a crucial task in contractor pre-qualification is to establish a set of decision criteria through which the capabilities of contractors are measured and judged. However, in the UK, there are no nation-wide standards or guidelines governing the selection of decision criteria for contractor pre-qualification (Ng and Skitmore, 1999, p. 607).

Holt et al. (1995), Hatush and Skitmore (1997a) and Kumaraswamy et al. (2005) had identified common criteria for pre-qualification and bid evaluation and proposed improved methodologies for contractor selection. Holt et al. (1995) proposed a quantitative multi-attribute based model, whereas Kumaraswamy et al. (2005) discussed contractor evaluation through an appraisal of inputs and assessment of outputs using 'feedforward' and 'feedbackward' approaches. Hatush and Skitmore (1997a) prescribed a Program Evaluation and Review Technique (PERT) based methodology for assessing and evaluating contractor data for the purposes of pre-qualification and bid evaluation.

To maximise the chances of successful projects construction clients should first consider and decide what represents quality to their organisations or to particular projects (APM, 2006, pp. 22–23). This criteria can then be used to compile a weighted list of requirements from which contractors can be judged on their respective attributes against the set values and a scoring matrix (Thomas and Thomas, 2005). However, Hatush and Skitmore (1997b, p. 129) argued that there is evidence to suggest that this approach is seldom adopted by most clients (and consultants) during pre-qualification selection processes. They concluded that many are 'more concerned with the process of retrieving completed proforma from candidate contractors than with under-taking any serious study of the relationships of this data to the project objectives'.

Ogunsemi and Aje (2006, p. 36) cited the seminal work of Russell and Skibniewski (1988) whom they credit as identifying five stages of pre-qualification criteria: 'references; reputation and past performance; financial stability; status of current work programmes; technical expertise; and project specific criteria'. In contrast, Hatush and Skitmore (1997a, p. 20) argued that the most common criteria considered by procurers during the pre-qualification and bid process are those pertaining to financial soundness, technical ability, management capability, and the health and safety performance of contractors. Gismondi (2017) highlighted the importance of following up with contractors' references, and asking them to provide a comprehensive post contract performance review of contractors. Gismondi also concluded that construction clients should follow up references from contractors' past clients not listed as referees.

Although there appears to be an overwhelming consensus that good quality pre-qualification processes have a positive correlation with improved project outcomes, a study by Wei *et al.* casts some doubt on this hypothesis:

> It seems reasonable to expect that high prequalification requirements will result in a competent contractor and consequently deliver a quality product in a timely manner and within budget, the findings of this study suggest that this is not necessarily true.
>
> (Wei *et al.*, 1999)

Wei *et al.* (1999) suggested that the one reason for this could be that construction clients who possess comprehensive pre-qualification practices tend also to have bureaucratic post award practices. These could stifle projects, causing delays through overcomplicated processes to deal with issues such as change management.

6.7 Pre-qualification models and methodologies

Ng and Skitmore (1999, p. 607) highlighted that despite Latham (1994) calling for standardisation of pre-qualification procedures, no one single model has been accepted and is widely used by the industry and that this is probably due to differing construction client objectives. However, there are many different models available to aid clients with contractor selection decisions. These include: multi-criteria model, fuzzy set model and the financial model (Darvish *et al.*, 2009, p. 611).

6.7.1 Multi-criteria model

The Latham Report (Latham, 1994) questioned the practice of selecting contractors based on the lowest price and suggested that a better approach would be to compare contractors based upon a range of attributes and deliverables. The multi-criteria model was developed to address this and involved ranking and weighting a range of client requirements and expectations against which contractors are compared (Jennings and Holt, 1998). Contractors are not solely judged on a single attribute, such as 'the financial offer' but several attributes are considered to arrive at a selection that represents the overall best value (Holt, 1998, p. 154). Latham confirmed that clients should base their choice of contractor on a value for money basis with proper weighting of selection criteria for skill, experience and previous performance, rather than accepting the lowest tender (Holt et al., 1995). Construction clients should be aware that multi-attribute decision making is defined by processes that involve designing the best alternative or selecting the best one from a set of alternatives, that has the most attractive overall attributes, and that involves the selection of the optimal alternative (Turskis, 2008, p. 225).

Morote and Francisco (2012, p. 9) were critical of the multi-attribute model claiming that it relies too much on the 'judgements and preferences of decision makers' and qualitative opinions and weightings which can be fuzzy and random. Some prominent researchers Ng and Skitmore (1999) and Hatush and Skitmore (1997b) endeavoured to address this randomness. Ng and Skitmore (1999, p. 609) established 35 pre-qualification criteria derived from their studies and questionnaires of client-side stakeholders (architects, project managers et al.) which were weighted based on a mean average of the respondents' responses regarding the importance rating for each criterion. Hatush and Skitmore (1997c) utilised Programme Evaluation and Review Technique (PERT) to assign weightings to criteria based on opinions of experienced practitioners, whom they asked to rank the likelihood of each criterion affecting time, cost or quality.

6.7.2 Fuzzy set model

Fuzzy sets were introduced by Zadeh in 1965 as an extension of the classical notion of sets. Fuzzy set theory is a mathematical theory in which elements have degrees of membership within a certain interval (Nasab and Ghamsarian, 2015).

Obtaining accurate and objective pre-qualification scores is often problematic as there are many inexact or qualitative criteria that are difficult to measure' (Yawe et al., 2005). Uncertainty and subjectivity in data collection during contractors' pre-qualification can result in ambiguous and obscure scoring as criterion such as reputation can be a grey area. Fuzzy set models use fuzzy set theory to improve the objectiveness of the assessment and results (Lam et al., 2001). A fuzzy logic system can be described as sorting linguistic qualitative data in a non-linear arrangement to compute numerical scales and sets. Fuzzy logic systems generally have four parts: fuzzifier, rules, inference engine and defuzzifier.

6.7.3 Financial model

One of the main checks that construction clients and consultants frequently utilise is a financial review to assess the suitability and capacity of contractors to undertake their projects. This involves the collection and scrutiny of contractors' financial statements and other information to establish if they have the economic capacity to carry out the work. Failure to carry out such checks could result in a contractor experiencing financial constraints and struggling to fund the regular progress of the works, or worse, they could become insolvent during the project (Huang et al., 2013, p. 254). There are a plethora of checks that can be carried out to ascertain contractors' financial standings, including their ability to obtain surety bonds (Awad and Fayek, 2012, p. 89), availability of working capital and documentary evidence of annual turnover levels (Doloi, 2009, p. 1249).

The two main financial reports used to assess the financial position of a company are the income statement and the balance sheet. The income statement describes the operational process over a period of time (usually one year) and reveals sales (revenues), costs, expenses, and profits or losses of a company over that time. The balance sheet is a snapshot of the financial position of a company at a particular point in time and describes its assets and liabilities. Net worth is the difference between what the company has and what it owes. Net worth is also helpful in determining the extent to which a company is leveraged. Leveraging refers to borrowing money to operate the company or to buy out shares in the company. Leverage measures indicate how much of the ability of the company to conduct business comes from borrowed funds (Pilateris and McCabe, 2003, p. 488).

6.8 Designing a new way for construction clients to select their contracting partners

Earlier in this chapter a number of practices were identified that could help to improve the chances of construction clients selecting reliable and proactive contracting partners. Conversely, a number of allegedly common practices were also reviewed that could be counter-productive to engendering collaborative relationships. The study by Taylor (2017) was conducted to identify and determine a new more appropriate method to collate, measure and analyse contractor data with a view to selecting construction partners. The study collected data from survey questionnaires and interviews based on the extent to which pre-qualification questionnaires are used and to assess project outcomes. This was to measure whether the quality of pre-qualification questionnaires used by construction clients influences project success. Examples of the questions used to measure each of these two variables are illustrated below:

Pre-qualification questions

> General question asking if detailed pre-qualification methods are carried out.
> Asks the respondent if the pre-qualification questionnaire includes a section relating to contractors' past performances.
> Asks if contractors' referees are contacted to corroborate contractors' past performances.

Asks if the past clients of contractors that were listed as referees are contacted to provide feedback on contractor's past performances.

Asks if official platforms such as Constructionline are contacted to obtain contractors' KPIs.

Asks if the respondent or the respondent's company are ever themselves contacted by official platforms such as Constructionline to rate contractors' performances.

Asks if the respondent or the respondent's company are ever themselves contacted by other clients or consultants to rate contractors' performances.

Asks if curricula vitae (CVs) are requested from contractors for proposed staff.

Asks if interviews of the contractor's proposed staff are carried out at pre-qualification stage.

Asks if the results of KPIs are ranked and threshold criteria used to disqualify applicants.

Asks if professional pre-qualification programmes, software or models are utilised to compare and select contractors.

Asks if financial health checks are carried out on contractors at pre-qualification stage, such as the one by Dun and Bradstreet.

Asks if BIM is used on projects.

Project outcome questions

Asks if contractors' stage two prices were generally fair and reasonable.

Asks if contractors completed projects on time.

Asks if contractors worked in an open and collaborative manner.

Asks if the contractors' stage two prices were within the original budget allowance set by the client.

Asks if the project risks were shared fairly and appropriately, inferring between the contractor and the client.

Asks if a disproportionate amount of 'construction' risks were borne by the contractor. The inference in the speech marks is to differentiate between general developers' risks and risks relating to the actual construction phase. This could also be loosely categorised as risks after construction has commenced.

Similar to the last question, this question asks if a disproportionate amount of 'construction' risks were borne by the client.

Asks if existing contractors are invited to tender for future competitive two stage tenders.

Asks if the quality of completed projects is generally finished to an acceptable standard.

Asks if repeat contracts were or likely to be negotiated with existing contractors.

Asks if the project out-turn costs are generally fair and reasonable.

> Asks if clients paid a premium for deterioration in the skills base of contractors' staff.
> Asks if the standard of health and safety management by the contractor was good.
> Asks if the contractor's sustainability ethos was good.
> Asks the respondent to rate the overall performance of contractors.

6.9 The quality of pre-qualification processes and their influence on project success

The overall results from the study by Taylor (2017) concluded a positive effect of 14 per cent correlation between pre-qualification practices and successful project outcomes, which is deemed to be a significant influence. Notwithstanding this relationship, not all construction clients carry out formal pre-qualification processes all of the time. Surprisingly only 8 of the 31 respondents stated that their company 'always carries out a detailed pre-qualification process before shortlisting contractors for two stage tenders'. This is quite an alarming revelation and something construction clients should address in 'turning the tide' and deploying best practices to support successful project outcomes. Accordingly, it would appear that significant improvements could possibly be made to increase the likelihood and the success rate of projects, by practitioners simply employing and carrying out more robust pre-qualification methods. This position is supported by one of the construction clients who was interviewed for this book:

> The more checks you do the better at pre-qualification stage to determine if a contractor is the right company to be given the opportunity and responsibility to tender for and more importantly successfully build your project!

A section in the pre-qualification questionnaires relating to past performance found a significant degree of correlation of 17 per cent against project outcomes. However, it appeared that the information that contractors provide to demonstrate past performance is not always corroborated. From the findings, contractors' referees were frequently contacted (70 per cent) to obtain first hand feedback on performance but past clients (not listed as referees) were contacted to obtain first hand feedback in just over half of cases (52 per cent). A further test on 'contacting contractor's referees' and project outcomes revealed a correlation of 17 per cent which again represents a significant influence. These correlation trends support the views of Gismondi (2017) who opined about the importance of following up with contractors' references. Even to most lay persons this must seem like a glaringly obvious and sensible course of action. How many people would invite a builder into their home, to carry out the most menial of projects, without first obtaining some kind of reference? Yet only 25 per cent of the respondents stated that they 'always' check out contractors' references.

The study by Taylor (2017) also aimed to determine and analyse if current pre-qualification practices are sufficient to allow clients to identify and select collaborative contractors. Unfortunately, only 35 per cent of respondents stated that contractors had mostly or always worked in a collaborative manner, with a modest positive correlation of 8 per cent between the overall pre-qualification scores and collaborative contractors. This was not a glowing endorsement that current pre-qualification practices are sufficient to allow clients to identify and select collaborative contractors. The results of some of the data led to a speculation that clients are overlooking this key soft skills attribute in their selection processes. Accordingly, construction clients should reflect on including more soft skills based criteria and analysis as part of their pre-qualification selection processes.

One of the reasons that use of Building Information Modelling (BIM) was included in the questionnaire was due to the discovery during the literature review that according to *Constructing Excellence* (BRE, 2018) this is one of the major practices to promote collaborative working. The results of the Taylor study appeared to corroborate this assertion showing a positive correlation (11 per cent) between BIM measures influencing collaboration. Unfortunately, despite BRE and the government's drive to promote BIM, its use is not wide in the private sector, as the total score for BIM use in the sample group was only 57 per cent.

It was interesting from Taylor's study that contractors who were deemed to be 'collaborative' were rewarded with repeat opportunities to tender for further work for construction clients. There was a moderately strong 16 per cent correlation between the two events. Conversely, it was disconcerting to observe a lack of any correlation coefficient between collaborative contractors and negotiated further work. It appeared that the construction client respondents did not show any inclination to reward such collaboration by negotiating future contracts. This could demonstrate an unnecessary burden of wasted resource and tendering costs for construction clients, which could be regarded as a *raison d'etre* of the collaborative working agenda. For construction clients this should be an area of careful reflection.

The third and final aim of Taylor's study was to determine and analyse the extent that construction clients are using objective assessments of KPIs to shortlist contractors for tenders. It was discovered that most researchers in this field claim that many construction clients used only intuition and subjective judgement to pre-qualify contractors for tenders. An interesting correlation was observed between the use of KPIs as a pre-qualification method and project outcomes. There was a moderately strong (13 per cent) correlation indicating that the use of KPIs increases the likelihood of achieving fair project out-turn costs. However, the results of the survey sample indicated that many construction clients could be failing to take advantage of this simple technique.

For the purposes of the assessment, the KPI category was classed as an objective process. It could, however, be construed that unless the threshold criteria has been prescribed as, say a company wide diktat, that this process is still a form of subjective assessment. In this case, the logical category or question to focus on would have been 'Professional pre-qualification programmes, software or models are utilised to compare and select contractors during pre-qualification assessments.'

6.10 Summary and conclusions

The following objectives have been discussed in this chapter and a compilation of the findings from Taylor's study have been summarised.

6.10.1 Determination of whether quality of pre-qualification processes by construction clients influences project success

It has been seen from Taylor's study that pre-qualification processes for construction clients do indeed influence project success. Accordingly, it follows that those construction clients that utilise a range of good quality diligent pre-qualification checks, have a greater chance that such diligence will yield positive results and projects completed in a manner that is considered as being successful. Despite this fact, many construction clients are failing to employ and make full use of all the pre-qualification tools available to them. In some instances they are not using them at all before shortlisting and appointing contractors for two-stage tenders, which could be affecting their chances of procuring successful projects.

Construction clients should develop and implement project specific and objective pre-qualification processes and ensure that any information submitted by contractors is validated by contacting cited referees, and if possible other past clients not identified by contractors.

6.10.2 Determination of whether construction clients' current pre-qualification practices are sufficient to identify and select collaborative contractors

The findings of the survey undertaken as part of Taylor's study showed only a modest correlation between the pre-qualification processes and 'collaborative contractors'. However, previous literature referred to earlier in this chapter clearly suggested that 'prior learning' about contractors' backgrounds and their *modus operandi* improves the prospects of encountering or engendering trust based collaborative relationships. Logically it should follow that to increase the likelihood of sourcing collaborative contractors, construction clients should utilise the wide spectrum of checks available to them during the investigatory pre-qualification stage. However, the findings show that many construction clients have failed to take full advantage of all available sources of information and tools, such as contacting past clients for references and the use of BIM.

It is clearly evident that the construction industry, certainly in the UK, still has an 'adversarial problem', and the findings of Taylor's study to some extent corroborate this. Adversarial approaches among main contractors are seemingly still a prevalent issue in the industry. However, the question arises, of how much of this is due to the lack of incentivisation from construction clients to contractors to be more collaborative; clients have still not fully embraced the concept of collaborative working themselves.

6.10.3 Determination of the extent to which construction clients are using objective assessments of KPIs to short-list contractors for tenders

According to Taylor's study, pre-qualification processes increase the likelihood of achieving successful project outcomes. Findings corroborated and showed that simple ranking and comparing of contractors' KPIs (with minimum set prescribed threshold criteria) improved the likelihood of projects achieving fair out-turn costs. Disappointingly only approximately a third of the construction client respondents stated that they either mostly or always used KPIs which could represent a non-committal approach in this case.

6.10.4 Potential recommendations and drivers for change and improvement

The initial aim of Taylor's study was to establish what information and methods are available to construction clients to help them to identify and select suitable contractors for long term collaborative relationships, and in doing so improve the chances of successful project outcomes. Irrespective of which pre-qualification methods are employed, the findings of this study showed that there are many tools and processes available to construction clients; and if used, increase the chances (regardless of the *modus operandi*) of achieving successful project outcomes. However, to improve the general standard and promote increased use, maybe the construction industry should develop and implement a National British Standard (NBS) for pre-qualification. This could be utilised by construction clients and assist them to remove much of the subjectivity and vagaries which are associated with current practices. Possibly future studies could consider the introduction of an industry wide contractor KPI data sharing platform for construction clients. It is acknowledged that this would require much research and consultation, especially regarding the obvious legal complexities involving defamation and associated litigation cases. If, however, the industry could reach a workable solution and implement a KPI data sharing platform, surely nothing would make a greater impact and allow construction to finally shake off its adversarial nature and embrace a future where excellence of service is rewarded with increased business opportunities.

References

APM (2006). *Association for Project Management: Body of Knowledge*. 5th ed. London: Association for Project Management.

Awad, A and Fayek, AR (2012). A decision support system for contractor prequalification for surety bonding. *Automation in Construction*, 21(1), 89–98.

Banner, D and Cooke, R (1984). Ethical dilemmas in performance appraisal. *Journal of Business Ethics*, 3(4), 327–333.

Banwell, H (1964). *The Placing and Management of Contracts for Building and Civil Engineering Work*. London: HMSO.

BBC (2018). *Carillion collapse raises job fears*. Available at: www.bbc.co.uk/news/business-42687032 [Accessed 17th January 2018].

Beales, R (2008). Call to make Constructionline mandatory. *Contract Journal*, 20 February, p. 3.

Beatham, S, Anumba, C and Tony, T (2004). KPIs: a critical appraisal of their use in construction. *An International Journal*, 11(1), 93–117.

Bouygues UK (2018). *Main Contractor Zone*. Available at: www.constructionline.co.uk/member-zone/main-contractors/ [Accessed 24th May 2018].

Bradshaw, J and Chang, S (2013). Past performance as an indicator of future performance: selecting an industry partner to maximize the probability of program success. *Defense AR Journal*, 20(1), 59–80.

BRE (2018). *Constructing Excellence – KPIs and Benchmarking*. Available at: constructingexcellence.org.uk/kpis-and-benchmarking/ [Accessed 18th May 2018].

Challender, J, Farrell, P and Sherratt, F (2014). *Partnering Practices: An Investigation of Influences on Project Success*. Available at: www.arcom.ac.uk/-docs/proceedings/ar2014-1039-1048_Challender_Farrell_Sherratt.pdf [Accessed 17th January 2018].

Construction News (2014). Contractors pledge support for pre-qual register. *Construction News*, 7 May.

Constructionline (2018). *PAS 91*. Available at: www.constructionline.co.uk/pas-91/ [Accessed 25th May 2018].

Costain (1964). *Banwell Committee (Report)*. Available at: hansard.millbanksystems.com/commons/1964/nov/09/banwell-committee-report [Accessed 22nd January 2018].

Cowper, J and Samuels, M (1996). *Performance Benchmarking in the Public Sector*. Available at: www.oecd.org/unitedkingdom/1902895.pdf [Accessed 11th May 2018].

Critchlow, J (1998). *Making Partnering Work in the Construction Industry*. Oxford: Chandos Publishing Limited.

Darvish, M, Yasaei, M and Saeedi, M (2009). Application of the graph theory and matrix methods to contractor ranking. *International Journal of Project Management*, 27(6), 610–619.

Doloi, H (2009). Analysis of pre-qualification criteria in contractor selection and their impacts on project success. *Construction Management and Economics*, 27(12), 1245–1263.

Dunton, J (2015). *Capita buys Constructionline for £35m*. Available at: www.building.co.uk/capita-buys-constructionline-for-£35m/5073467.article [Accessed 24th May 2018].

EC Harris LLP (2013). *Supply Chain Analysis into the Construction Industry: A Report for the Construction Industry Strategy*. Available at: www.gov.uk/government/publications/construction-industry-supply-chain-analysis [Accessed 21st May 2018].

Egan, J (1998). *Rethinking Construction: The Report of the Construction Task Force*. London: DETR.TSO. 18–20.

Farmer, M (2016). *The Farmer Review of the UK Construction Labour Model: Modernise or Die*. London: Construction Leadership Council (CLC).

Farmer, M (2017). Modernise or die: one year on. *Building*, 13 October, p. 33.

Fernie, S, Leiringer, R and Thorpe, T (2006). Change in construction: a critical perspective. *Building Research and Information*, 34(2), 91–103.

Fong, P and Choi, S (2000). Final contractor selection using the analytical hierarchy process. *Construction Management and Economics*, 18(5), 547–557.

Ghaffarianhoseini, A et al. (2017). Building Information Modelling (BIM) uptake: clear benefits, understanding its implementation, risks and challenges. *Renewable and Sustainable Energy Reviews*, 75, 1046–1053.

Gismondi, A (2017). Challenges with prequalification process addressed during panel. *Daily Commercial News*, 16 June, pp. 1–2.

Gulf Construction: Al Hilal Publishing & Marketing Group (2008). Two-stage tendering is a useful option. *ProQuest*, 1 December, p. 1.

Hatush, Z and Skitmore, M (1997a). Criteria for contractor selection. *Construction Management and Economics*, 15(1), 19–38.

Hatush, Z and Skitmore, M (1997b). Evaluating contractor prequalification data: selection criteria and project success factors. *Construction Management and Economics*, 15(2), 129–147.

Hatush, Z and Skitmore, M (1997c). Assessment and evaluation of contractor data against client goals using PERT approach. *Construction Management and Economics*, 15(5), 327–340.

HM Government (2013). *Construction 2025 – Industry Strategy: Government and Industry in Partnership*. London: HM Government.

Holt, G (1998). Which contractor selection methodology?. *International Journal of Project Management*, 16(3), 153–164.

Holt, GD, Olomolaiye, P and Harris, F (1995). A review of contractor selection practice in the UK construction industry. *Building and Environment*, 30(4), 553–561.

Huang, Y and Wilkinson, I (2013). The dynamics and evolution of trust in business relationships. *Industrial Marketing Management*, 42, 455–465.

Huang, W et al. (2013). Contractor financial prequalification using simulation method based on cash flow model. *Automation in Construction*, 35(11), 254–262.

Jennings, P and Holt, G (1998). Prequalification and multi-criteria selection: a measure of contractors' opinions. *Construction Management and Economics*, 16(6), 651–660.

Kumaraswamy, MM, Yean, FYL, Rahman, MM and Phng, ST (2005). Constructing relationally integrated teams. *Journal of Construction Engineering and Management*, 131(10): 1076–1084.

Kwakye, AA (1991). *Fast Track Construction*. Ascot: The Chartered Institute of Building.

Lam, KC et al. (2001). A fuzzy neural network approach for contractor prequalification. *Construction Management and Economics*, 19(2), 175–188.

Laryea, S and Hughes, W (2008). How contractors price risk in bids: theory and practice. *Construction Management and Economics*, 26(9), 911–924.

Latham, M (1994). *The Latham Report*. Available at: constructingexcellence.org.uk/wp-content/uploads/2014/10/Constructing-the-team-The-Latham-Report.pdf [Accessed 29th November 2017].

Leflty, M (2014). Too much information? One day the state might be selling the family data. *The Independent*, 25 July, p. 56.

Lehtiranta, L (2014). Risk perceptions and approaches in multi-organizations. *International Journal of Project Management*, 32, 640–652.

Lindkvist, C (2015). Contextualizing learning approaches which shape BIM for maintenance. *Built Environment Project and Asset Management*, 5(3), 318–330.

Luu, V, Kim, S and Huynh, T (2008). Improving project management performance of large contractors using benchmarking approach. *International Journal of Project Management*, 26, 758–769.

Mann, W (2014). Government to sell off Constructionline. *New Civil Engineer*, 15 July.

McCabe, S (2001). *Benchmarking in Construction*. 1st ed. London: Blackwell Science Ltd.

McDonald, T (2013). *Collaborative Risk Management.* Available at: www.fgould.com/worldwide/articles/collaborative-risk-management/ [Accessed 22nd May 2018].

Morote, A and Francisco, R-V (2012). A fuzzy multi-criteria decision making model for construction contractor prequalification. *Automation in Construction*, 25(4), 8–19.

Naoum, SG and Egbu, C (2016). Modern selection criteria for procurement methods in construction: a state-of-the-art literature review and a survey. *International Journal of Managing Projects in Business*, 9(2), 309–336.

Nasab, HH and Ghamsarian, MM (2015). A fuzzy multiple-criteria decision-making model for contractor prequalification. *Journal of Decisions System*, 24(4), 433–444.

National Joint Consultative Committee for Building (1996). *Code of Procedure for Two Stage Selective Tendering.* London: NJCC Publications.

Ng, S and Skitmore, R (1999). Client and consultant perspectives of prequalification criteria. *Building and Environment*, 34(5), 607–621.

Ng, S, Palaneeswaran, E and Kumaraswamy, M (2002). A dynamic e-Reporting system for contractor's performance appraisal. *Advances in Engineering Software*, 33(7), 339–349.

Ogunsemi, D and Aje, IO (2006). A model for contractors' selection in Nigeria. *Journal of Financial Management of Property and Construction*, 11(1), 33–34.

Osipova, E and Eriksson, P (2011). How procurement options influence risk management in construction projects. *Construction Management and Economics*, 29(11), 1149–1158.

Osipova, E and Eriksson, P (2013). Balancing control and flexibility in joint risk management: lessons learned from two construction projects. *International Journal of Project Management*, 31, 391–399.

Palaneeswaran, E and Kumaraswamy, M (2001). Recent advances and proposed improvements in contractor prequalification methodologies. *Building and Environment*, 36, 73–87.

Peace, S (2008). *Partnering in Construction.* Salford: Construction Managers' Library.

Pilateris, P and McCabe, B (2003). Contractor financial evaluation model (CFEM). *Canadian Journal of Civil Engineering*, 30(3), 487–499.

Procurement Group (1999). *Procurement Guidance No 4: Teamworking, Partnering and Incentives.* London: HM Treasury.

Rawlinson, F and Farrell, P (2010). UK construction industry site health and safety management: an examination of promotional web material as an indicator of current direction. *Construction Innovation*, 10(4), 435–446.

Raymond, J (2008). Benchmarking in public procurement. *An International Journal*, 15(6), 782–793.

Reed, J (2018). *When should a contractor contest a CPARS rating?.* Available at: www.mwllegal.com/2018/01/18/when-should-a-contractor-contest-a-cpars-rating/ [Accessed 21st May 2018].

Reeve, P (2013). It's time to standardise in construction. *Construction News*, 22 November.

Rigby, J, Dewick, P, Courtney, R and Gee, S (2014). Limits to the implementation of benchmarking through KPIs in UK construction policy. *Public Management Review*, 16(6), 782–806.

Saad, M and Patel, B (2006). An investigation of supply chain performance measurement in the Indian automotive sector. *Benchmarking: An International Journal*, 13(1), 36–53.

Steele, A, Todd, S and Sodhi, D (2003). Constructionline: a review of current issues and future potential. *Structural Survey*, 21(1), 16–21.

Sutton (2008). Call to make Constructionline mandatory. *Contract Journal*, 20 February, p. 3.

Taylor, S (2017). *Prequalification practices and project outcomes in connection with two stage tendering*. Unpublished MSc Construction Management dissertation. University of Bolton, UK.

The KPI Working Group (2000). *KPI Report for the Minister for Construction*. Available at: assets.publishing.service.gov.uk/government/uploads/system/uploads/attachment_data/file/16323/file16441.pdf [Accessed 21st May 2018].

Thomas, G and Thomas, M (2005). *Construction Partnering and Integrated Teamworking*. 1st ed. Pondicherry: Blackwell Publishing Ltd.

Turskis, Z (2008). Multi-attribute contractors ranking method by applying ordering of feasible alternatives of solutions in terms of preferability technique. *Technological and Economic Development of Economy*, 14(2), 224–239.

Vass, S and Gustavsson, T (2017). Challenges when implementing BIM for industry change. *Construction Management and Economics*, 35(10), 597–610.

Wei, L, Krizek, R and Hadavi, A (1999). Effects of high prequalification requirements. *Construction Management and Economics*, 17(5), 603–612.

Wolstenholme, A (2009). *Never Waste a Good Crisis; A Review of Progress Since Rethinking Construction and Thought for Our Future*. London: Constructing Excellence in the Built Environment. Available at: http://dspace.lboro.ac.uk/dsspacejspui/bitstream/2134/6040/1/Wolstenholme%20Report%20Oct%20Oct%202010.pdf [Accessed 13th January 2014].

Yawe, L, Shouyu, C and Xiangtian, N (2005). Fuzzy pattern recognition approach to construction contractor selection. *Fuzzy Optimization and Decision Making*, 4(2).

7 The 'Intelligent Client'

A model of procurement built on relationship management between construction clients and the supply chain

7.1 Introduction

This chapter, based around a research study by Challender (2017), and is focused on the importance for construction clients to build the trust of project teams through collaborative strategies. The research was based on both quantitative and qualitative methodologies and involved questionnaire survey and interviews conducted on construction clients in many different sectors. This mixed method approach was adopted for the research to triangulate data collection with the aim of increasing validity of its findings. This study stemmed from 'The Intelligent Client' (HM Government, 2012) which recommended a model of procurement and relationship management between construction clients and the supply chain. This creates a framework in the first place for a much more sophisticated approach. The chapter also explores the extent to which construction clients can utilise the theory of trust as a viable tool in collaboratively procuring more successful construction outcomes.

The study findings in this chapter are felt relevant for construction clients as they give an understanding of how trust building mechanisms can be designed and implemented for improving project outcomes. A quantitative study was carried out from survey questionnaires and by subjecting data to correlation analysis. The research population was restricted to those contracting, consulting and client organisations that have had experience of collaboratively procured projects.

7.2 What do we mean by collaborative strategies?

The terms 'partnering' and 'collaborative working' are used interchangeably within this chapter, referencing a wider philosophy of trust, fairness and equity, rather than specific details of practice. Although the concept of these terms is widely used, there appears to be a distinct lack of understanding of what it means to collaborate and the ways and means to improve and support its successful practice (Patel *et al.*, 2012, pp. 1–4). The Construction Industry Council defines partnering as 'a structured management approach to facilitate team working across contractual boundaries' (Construction Industry Council, 2016). The research was done predominantly to educate construction clients about the benefits of collaboratively procured contracts and trust building mechanisms. These have long been advocated by government funding bodies in a concerted attempt to improve successful outcomes for construction projects.

7.3 An awareness for construction clients on issues around collaborative procurement strategies and trust

Over recent years there have been many arguments presented that traditional procurement strategies have achieved low client satisfaction levels, poor cost predictability and time certainty largely attributable to coordination difficulties associated with separation of design and construction and the greater need for teamwork (Latham, 1994, pp. 81–83; Egan, 1998, pp. 18–21; Egan, 2002, p. 6). Collaborative working at an early stage between contractors and design teams has been, post-Latham, regarded as a means to bridge the gap between design and construction to improve project outcomes. Accordingly, many have identified collaborative procurement routes such as partnering as a critical success factor on construction projects (Vaaland, 2004). Despite the aforementioned arguments in support of collaborative procurement strategies, such approaches have attracted their critics. There are, for instance, views that partnering practices within the UK construction industry have failed to realise the full extent of benefits and positive effects that they have experienced in sectors such as manufacturing (Miller et al., 2002; Morgan, 2009, pp. 7–9; Winch, 2000 as cited in Bygballe et al., 2010, pp. 240–245). This was reiterated by Gadde and Dubois (2010, p. 262) who stated simply that 'partnering has not lived up to expectations'. Perhaps this could partly explain the reported downward trend in popularity and participation of what was heralded as a major breakthrough in construction management and a rather unwelcome return to traditional procurement based on lowest cost tenders (RICS, 2012).

There is a wide ranging consensus that potential barriers in the construction industry could have hindered successful partnering which include fear of the unknown, perceived loss of control, uncertainty and the lack of understanding of how to change the way one works (Thurairajah et al., 2006, pp. 1–4). Within the context of these potential barriers, some participants of the study carried out by Challender (2017) may have been displaying acted behaviours to collaborative working but ultimately behaving in a consistent way to their true beliefs. Accordingly, they may only 'pay lip service to the principles of partnering' (Gadde and Dubois, 2010, pp. 260–262). Certainly Boes and Doree (2013) corroborate this view and their findings indicate that partnering can frequently be classified as superficial and be used to disguise traditional adversarial power relationships and attitudes. There is, however, a careful balance required, in and around these arguments, rather like competitiveness versus cooperation, in how construction related organisations behave, linked very much to predictability and confidence in the other party brought about by trust (Cheung et al., 2001). Perhaps the aforementioned attitudes and behaviours demonstrate that partnering principles and philosophies are still not being fully embraced by the construction industry.

Over recent years, organisations have largely focused on increasing partnering strategies for collaborative procurement of major capital projects. Such initiatives are often heralded as vehicles to obtain best value, improve levels of quality and optimise service delivery. Yet there is still evidence of low levels of client satisfaction, owing mostly to poor cost and time predictability, which have in turn been attributed to low level of trust in practice (Kumaraswamy and Mathews, 2000; Kumaraswamy et al., 2005; Chow et al., 2012). This potential lack of trust in collaborative working practices, alongside the general recent 'tightening' of the market, could possibly explain the downward trend in favour of more market based approaches to construction procurement (Ross, 2011).

7.4 Trust as a collaborative necessity for construction clients

In previous studies of collaborative working, very little attention has been focused on the trust building process (Harris and Lyon, 2013). Support for this argument comes from Thorgren et al. (2011) who concurred that 'scant attention has been paid to the role and development of trust in partner alliances'. The study by Challender (2017) therefore sought to fill this gap in knowledge by examining trust building attributes and mechanisms, and exploring the influence of these on generating trust in construction partnering. It was also designed to address calls for greater insight into how trust is created, mobilised and developed (Huemer, 2004) and for more understanding of the effects and impact of other factors interacting with trust (Huang and Wilkinson, 2013).

Trust has been defined as an 'expectancy held by an individual that the word, promise, oral or written statement of another individual or group can be relied upon' (Rotter, 1980) or conversely as a 'belief in a person's competency' (Sitkin and Roth, 1993). The *Oxford Current English Dictionary* (1990) offered a further definition as 'confidence in or reliance on some quality; attribute of a person and thing; or the truth of a statement'. Conversely others have regarded trust as 'an expression of confidence that cannot be compromised by the actions of another party' (Jones and George, 1998) or 'one party's optimistic expectation of the behaviour of another, when the party must make a decision about how to act under conditions of vulnerability and dependence' (Hosmer, 1995). Perhaps the most prolific definition in recent times comes from Mayer et al. (1995) who described it as 'the willingness of a party to be vulnerable to the actions of another party based on the expectation that the other party will perform a particular action important to the trustor, irrespective of the ability to monitor or control that other party'. This definition is supported by Bigley and Pearce (1998) who suggested that for trust to be exercised, it requires a degree of vulnerability where at least one of the partners has potentially something important at stake and where there is potential for betrayal and opportunistic behaviours.

Much has been written on trust as a collaborative necessity (Larson, 1997; Chan et al., 2004, p. 230; Walker, 2009; Morrell, 2011) and this has largely focused on the advantages and merits of collaborative working and practice. Findings from such studies have indicated that greater levels of trust and collaborative

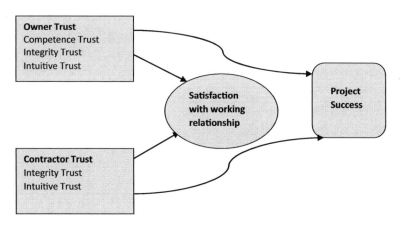

Figure 7.1 Conceptual framework between trust and success

working can increase the successful performance of projects and this is illustrated in Figure 7.1.

Notwithstanding this, there is still limited research to investigate and analyse the specific impact of trust within partnering practices (Wu and Udeaja, 2008). Furthermore, according to Pinto *et al.* (2009), there has also been an overwhelming lack of empirical studies with regard to building collaborative trust in procurement strategies and the influence of certain critical factors or constructs on collaborative trust. Such deficiencies have become apparent despite findings from Adams (2009) which indicated that the overall levels of trust constructs and attributes influence the level of 'trust intensity'. Trust constructs in this sense are 'the foundation or production processes upon which trust intensity can be established' and trust attributes relate to situations and circumstances where trust can be developed (McKnight *et al.*, 1998, Kadefors, 2004; as cited in Adams, 2009).

The model is supported by findings by Fukuyama (1995, as cited in O'Hara, 2006), which advocated that client/consultant/contractor relationships which generate trust will flourish and bring with them economic and social advantages, beneficial for organisations and society at large. It is the understanding of trust building mechanisms around improving partnering procurement strategies at the start of the process that this book will give particular attention to. This is regarded as an important contribution given that previous studies have identified the lack of trust as the main deficiency and 'root problem' for collaboratively procured projects in the UK.

'The Intelligent Client' (HM Government, 2012) recommended a model of procurement through which relationship management between the client and the supply chain creates a work environment in which collaboration and trust flourishes.

Whilst findings from such studies have indicated that greater levels of collaborative working can increase the successful performance of projects (Wong *et al.*,

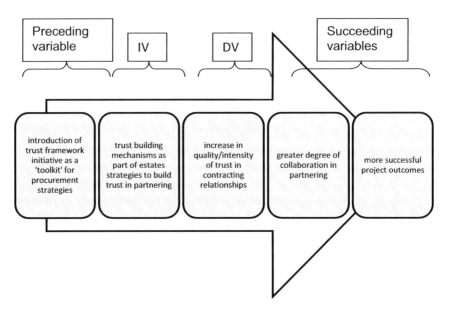

Figure 7.2 Flowchart to illustrate the influence of trust on improved project performance

2008; Pinto et al., 2009), there is still limited research to investigate and analyse the specific impact of trust within partnering practices (Wu and Udeaja, 2008). To address this deficiency, construction clients should focus 'upstream' on those constructs, attributes, factors, mechanisms and initiatives which could influence trust in the context of partnering practices. To assist in this pursuit, the study by Challender (2017) identified and evaluated trust 'generators', referred to as 'trust building mechanisms', and 'inhibitors' or barriers in this respect. This is designed to facilitate greater understanding of how trust building initiatives can be designed and implemented in developing a framework for improving public sector procurement strategies. The flowchart in Figure 7.2 illustrates the variables (independent and dependent) and the relationships between trust, collaboration and performance in securing more successful project outcomes.

The study by Challender (2017) identified previous research relating to the influence of key critical factors on improving project performance and shaping best practice. Despite this, very little has been written on how these same factors can influence trust, especially in partnering arrangements. Such success factors for influencing construction performance generally include ethical (Vee and Skitmore, 2003; Liu et al., 2004; Rahman et al., 2007; Walker, 2009; RICS, 2010), economic (Walker, 2009), motivational (Olomolaiye and Price, 1989, pp. 280–284; Steers et al., 1996; Carnall, 1999; Tabassi and Bakar, 2009, pp. 472–474; Tabassi et al., 2011) and organisational considerations (Thurairajah et al., 2006, p. 8; Bresnen and Marshall, 2010).

According to Ceric (2014) the issue of trust, in the quest for greater commitment to partnering, needs to be fully addressed by the construction industry to facilitate better working relationships. Mistrust and scepticism towards partnering practices could, however, be the result of a general lack of understanding of partnering philosophies and the wider benefits of collaborative working (Dainty et al., 2001). To address these issues, a greater awareness of the underlying problems of trust relating to adversarial behaviours is required by construction clients and passed down through their project teams. Furthermore, there is an argument that the formation stages of partnering arrangements are commonly regarded as being particularly challenging, when negotiations are progressing and interactions between the parties are complex. At this point in time trust is considered to be a 'bonding agent' between collaborating partners and regarded as an 'essential foundation for creating relational exchange' (Langfield-Smith, 2008). In this way it can also be regarded as a mechanism for governance of partnerships which seeks to deter cheating and opportunism, especially where dependence is strong and uncertainty is great (Silva et al., 2012). Irrespective of the stage of its development, however, construction clients should be aware that trust in partnering can create the willingness for further collaboration and organisational alignment. This can also reduce the requirements for reducing the cost and time in managing the behaviours and actions of individuals (Walker and Hampson, 2003; Rowlinson and Cheung, 2004).

7.5 Potential benefits of trust for construction clients; incentives to trust

The construction industry has in the past been characterised by complex processes and exchanges of information, and some would believe that these can lead to the emergence of opportunistic behaviours. To address these issues, research has focused on theories relating to the creation and development of trust as a potential means to reduce opportunism (Chow et al., 2012). For example, the existence of trust has been an important counter weight when business environments and organisational relationships have been prone to hidden agendas and conflicting objectives (Silva et al., 2012). Other theories, conversely, advocated that trust within relationships can safeguard against excessive formal contractual relationships developing between partnering organisations (Pinto et al., 2009). This is supported from the perspective that it is not practical to include contractual provisions for every possible occurrence on projects, and that too many provisions could be misinterpreted as signs of mistrust (Kramer, 1999; Li, 2008). The same argument is supported by Colquitt et al. (2007) who found that the potential benefits of developing and nurturing trust in the workplace can have positive influences on job performance whilst allowing vital risk taking, where there are no other safeguards to protect partners. The emergence of these views would appear to justify claims that financial benefits can emanate from trusting relationships. Ceric (2014) certainly supports this argument and advocates that 'trust reduces agency costs both before and after the contract is signed between the principal and the agent'. Conversely, and perhaps unsurprisingly, the opposite effect is evidenced for mistrust in the same study with additional agency costs being generated.

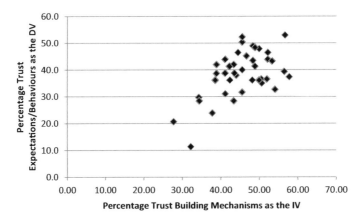

Figure 7.3 Scatter diagram illustrating the influence of trust building mechanisms (independent variable) to the level of trust generated on projects (dependent variable)

7.6 Research findings and discussion

7.6.1 Quantitative analysis: questionnaire data

The measured data relating to the survey questionnaires revealed a very strong correlation between the degree of trust building mechanisms embedded in projects and the degree of trust that is generated. This relationship between the two variables is illustrated as a scatter diagram in Figure 7.3. This is an interesting finding for construction clients especially as there is evidence that trust generated on projects improves project outcomes (Pinto et al., 2009).

7.6.2 Qualitative analysis: interview data

The qualitative analyses from interviews with eight participants revealed contrasting opinions on most of the trust building mechanisms when examined against previous literature sources.

When reflecting on which trust building mechanisms are more effective than others the qualitative analysis reveals that the following initiatives are considered to have greater influence on generating trust in partnering arrangements:

- Facilitation of regular workshops and review meetings at both executive and project team levels, specifically designed to resolve conflicts and problems.
- Formulation of strategies to develop mutually aligned corporate and strategic objectives between partnering organisations.
- Implementation of performance related 'gain share/pain share' partnering initiatives that are considered fair and equitable to partners.

- Implementation of transparent joint 'open book' financial accounting policies for all contractor/subcontractor valuations.
- Formulation of good internal and external communication strategies with frequent newsletters and email updates designed to keep all staff fully informed of project status.
- Adoption of democratic management styles between partners.
- Use of building contracts with terms and conditions that reflect a fair and equitable balance of power, risk and reward between partners.
- Deployment of resources between partners which are considered fair and equitable.
- Introduction and attendance of project teams at CPD workshops designed to support partnering principles, behaviours and attitudes.
- Formulations of strategies to ensure that senior management are involved in key decision making at various stages of projects.
- Introduction and adoption of joint recruitment policies and selection processes to include transparent joint 'open book' tender evaluation policies for all subcontractor/supplier appointments.
- Commitment of senior management in resourcing projects and to support key decision making.
- Implementation of joint management systems such as 'Business Collaborator' to facilitate greater transparency and sharing of information between partners.
- Introduction of joint relationship management policies to resolve any potential disputes, conflicts and difficulties raised by partners include 'issue resolution' processes.

These results broadly underpin the quantitative analysis, using correlation analysis, which revealed that these initiatives showed a strong influence over the amount of trust they generated.

From the interviews, workshops were regarded as being particularly effective when 'end users' from client organisations are involved and where there is open and free flowing dialogue around risk management. Furthermore performance related 'gain share/pain share' incentives were seen by participants as requiring robust and transparent key performance indicators. These should be easily and objectively measured against established agreed benchmarks to avoid disputes arising. An example might include financial incentives being based on savings achieved against target cost plans. Other forms of incentives could be effective through staff recognition schemes and as part of 'Investors in People' initiatives. Although securing the commitment of senior management is regarded as a highly effective trust building mechanism this is predicated on client executives embracing partnering philosophies. Where such senior decision makers advocate lowest price tendering in the guise of obtaining best value for governance adherence it can have a negative effect on trust and represents a major barrier for partnering strategies in such cases. Construction clients should take heed of this and avoid strategies linked to accepting lowest price tenders. Furthermore there were some trust building mechanisms that are considered less effective at generating trust in partnering arrangements and these included:

- Social functions and network events for project teams.
- Introduction of partnering charters (an example of which is included in Appendix 1C).
- Implementation of policies to ensure that only individuals who have membership of professional bodies are selected for project teams.

These results broadly underpin the quantitative analysis, using correlation analysis, which revealed that these initiatives showed only a weak to moderate influence over the amount of trust they generated. The findings could indicate that practitioners with limited resources in terms of time and cost should focus more on those initiatives that are deemed to have a stronger influence on trust. Partnering charters, for instance, could be generally so vague and meaningless in terms of describing the project objectives and behaviours, that it is perhaps not surprising that the research found participants not that enthusiastic about them. However, workshops could be regarded as creating the right environment for individuals to properly collaborate, share ideas and collectively solve problems, so perhaps it is not surprising that they were regarded as more effective trust building initiatives. In light of these considerations, it can be suggested that organisations should focus more on strategic initiatives linked with improvements in communications, incentives, fairness, information sharing and swift conflict resolution processes. Such mechanisms could become part of a 'partnering toolkit', geared to raising trust levels between project partners. Conversely, the findings also suggest that there should be less emphasis on arranging social events, preparing partnering charters and adopting restrictive recruitment policies around membership of professional bodies. Strategic partnering was felt to offer more beneficial outcomes and be more conducive to procuring successful collaborative outcomes than project specific partnering. This was explained by the perceived willingness to invest more in resources and based on the expected longevity of future relationships and work streams in the former case.

The overarching consensus emanating from the study by Challender (2017) supports the notion that trust building strategies play an extremely important role in influencing the levels and quality of trust in partnering. Accordingly construction clients, as leaders and 'project sponsors' should wherever possible pursue such initiatives to have a beneficial influence on project success. However, the level of influence of such mechanisms is dependent on the suitability and adaptability of different project types to partnering. Complex projects of longer duration are found to give more scope and opportunities for trust development within project teams.

7.7 Implications for construction clients in managing requirements and expectations for collaboration and trust

When considering the extent to which clients are prepared and commit to collaborating and trusting their construction partners it is important to set down

Table 7.1 Collaboration outputs, outcomes, key activities and expectations of clients and their construction partners

Project title obscured for confidentiality: Collaboration Champion outputs, outcomes, key activities and expectations	
Purpose	To facilitate the development and sustained application of effective collaboration processes and behaviour throughout the life cycle of the project.
Key customers of this role	Names removed for confidentiality reasons.
Outputs	Efficient and effective meetings. Improved collaboration skills within the project. Effective relationships based on generous listening, candid conversations and collaborative problem solving. A paper on effective collaboration practices and behaviour in construction projects.
Outcomes	An exemplar project for the Masterplan. A legacy of improved collaboration across Directorates and industry integration and partnerships. A model of best procurement practice. Industry acknowledgement as the exemplar for collaboration practices and behaviour in construction projects. Recommendations and repeat work through strategic partnering.
Key activities	To develop and enhance key skills for collaborative working. Facilitating the following 'creation' events: • Risk Management workshop. • Communications and Issues Resolution workshop. • Performance Measurement workshop. • Project Management and Systems workshops. Facilitating the following review and improvement events: • Core Group six weekly meetings. • Monthly Risk Management meetings. • Monthly communications and Issues Resolution meetings. • Monthly continu.ous improvement meetings. • Supporting the development and maintenance of effective relationships within the project. • Measuring and reviewing if collaboration is working effectively and recommending and initiating improvements. • Checking and reminding people on their agreed collaboration action points.

(continued)

Table 7.1 (Cont.)

Project title obscured for confidentiality: Collaboration Champion outputs, outcomes, key activities and expectations	
Client expectations of Partnering Champion in this role	Maintain confidentiality and keep things within the project. Be independent and not take sides. Provide frank feedback. Guide and bring innovative ideas. Establish relationships. Manage meetings. Oversee and manage the Trust Inventory Questionnaire quarterly with follow up conversations to identify blockages and recommend actions to reduce or remove them.
Partnering Champion's expectations of clients	Be receptive to frank feedback and avoid being defensive. Provide reasonable notice as much as possible when needing support. Agree a timetable of scheduled meetings that Collaboration Champion needs to facilitate. Respect Collaboration Champion role and make time to meet with this individual for planning and review purposes. Be open and candid at all times.

expectations to avoid any ambiguity. To enable such a commitment, it would be normal to agree those expectations amongst the whole project team, 'in the spirit and philosophy' of working together. This would normally be facilitated through a group workshop where roles, responsibilities, collaboration initiatives can be discussed and agree democratically. The example in Table 7.1 could illustrate how such commitments are documented which would cover such aspects as collaboration outputs, outcomes, key activities and expectations of clients and their construction partners.

These commitments could become developed and compiled into a 'partnering charter' which would act as a memorandum of understanding between the parties, covering behaviours, team working and approaches to partnering. An example of such a partnering charter is included in Appendix 1B.

This partnering charter would not normally be incorporated into the contract documents, as most probably would not be legally binding, but be displayed on construction sites, in prominent areas to encourage compliance and a spirit of team working.

7.8 Conclusions and recommendations

This chapter, through the research findings of Challender (2017), determined that there is an apparent lack of trust in project procurement strategies. Accordingly, construction clients should be aware of this major obstacle and consider strategies linked to partnering and building trust. In addressing this challenging dilemma construction clients should consider those factors which could influence trust building and which ones they can feasibly embed within projects. This will enable them to have a greater understanding of those trust building mechanisms that are potentially effective in 'turning the tide' and embedding more trust in their projects. Furthermore, trust building strategies have been established by this study as having a strong influence on raising trust levels on projects, especially when linked with partnering procurement. Examples of successful initiatives for construction clients could include strategies around incentive provisions, workshops, CPD, collaboration management systems, senior management commitment, open and joint evaluation policies and improved communications. Such measures or mechanisms could be designed by construction clients to increase the low levels of trust that exist on projects in pursuit of more successful project outcomes for their respective organisations. They are, however, heavily reliant on establishing mutually aligned corporate objectives between partnering organisations. Furthermore, they may provide the catalyst that 'keeps the partnering trust flag flying' in this regard, especially in those sectors which have seen a reduction in this procurement approach for construction projects in recent years.

This chapter of the book has identified barriers and obstacles to trust generation within partnering strategies. These mostly revolve around commercial issues and traditional adversarial behaviours of project teams. Improvement measures to address such issues include more focus on collaborative working as part of construction management curriculum, awareness of benefits of partnering philosophies through CPD and use of more collaborative management tools. Perhaps the biggest challenges for construction clients remain around culture change within the construction industry and seeking longer term collaborative relationships between their partnering organisations.

References

Adams, K (2009). *An examination of trust formation in the construction industry*. MSc Construction Management dissertation. University of Bolton, UK.

Bigley, GA and Pearce, JL (1998). Straining for shared meaning in organisation science: problems of trust and distrust. *Academy of Management. The Academy of Management Review*, 23(3), 405–421.

Boes, H and Doree, A (2013). Public procurement at local level in the Netherlands: towards a better client-contractor cooperation in a competitive environment. In Smith, SD and Ahiaga-Dagbui, DD (Eds), *Procs 29th Annual ARCOM Conference*, 2–4 September 2013. Reading, UK: Association of Researchers in Construction Management. 717–727.

Bresnen, M and Marshall, N (2010). Building partnerships: case studies of client-contractor collaboration in the UK construction industry. *Construction Management and Economics*. Available at: www.tandfonline.com/loi/rcme20 [Accessed 1st September 2013].

Bygballe, LE, Jahre, M and Sward, A (2010). Partnering relationships in construction: a literature review. *Journal of Purchasing and Supply Management*, 16, 239–253.

Carnall, CA (1999). *Managing Change in Organisations*. 3rd ed. Hertfordshire: Prentice Hall Europe.

Ceric, A (2014). Communication risk and trust in construction projects: a framework for interdisciplinary research. In Raiden, AB and Aboagye-Nimo, E (Eds), *Procs 30th Annual ARCOM Conference, 1–3 September 2014*. Portsmouth, UK: Association of Researchers in Construction Management. 835–844.

Challender, J (2017). *Collaborative trust in UK further education procurement strategies*. Unpublished PhD thesis. University of Bolton, UK.

Chan, APC, Chan, DWM, Chiang, Y, Tang, B, Chan, EHW and Ho, KSK (2004). Exploring critical success factors for partnering in construction projects. *Journal of Construction Engineering and Management*, 130(2), 188–189.

Cheung, SO, Lam, TI, Leung, MY and Wan, YW (2001). An analytical hierarchy process based procurement selection method. *Construction Management and Economics*, 19(4), 427–437.

Chow, PT, Cheung, SO and Chan, KY (2012). Trust-building in construction contracting: mechanism and expectation. *International Journal of Project Management*, 30(2012), 927–937.

Colquitt, JA, Scott, BA and LePine, JA (2007). Trust, trustworthiness, and trust propensity: a meta-analytic test of their unique relationships with risk taking and job performance. *Journal of Applied Psychology*, 92(4), 909–927.

Construction Industry Council (2016). *A Guide to Project Team Partnering*. Available at: www.cic.org.uk [Accessed 6th December 2016].

Dainty, A, Briscoe, G and Millet, J (2001). Subcontractor perspectives on supply chain alliances. *Construction Management and Economics*, 19(2), 841–848.

Egan, J (1998). *Rethinking Construction: The Report of the Construction Task Force*. London: DETR.TSO. 18–20.

Egan, J (2002). *Accelerating Change: Rethinking Construction*. London: Strategic Forum for Construction.

Gadde, LE and Dubois, A (2010). Partnering in the construction industry: problems and opportunities. *Journal of Purchasing and Supply Management*, 16(4), 254–263.

Harris, F and Lyon, F (2013). Trans-disciplinary environmental research: building trust across professional cultures. *Environmental Science and Policy*, 31(2013), 109–119.

HM Government (2012). *Government Construction Strategy: Final Report to Government by the Procurement Client Task Group*. London: HM Government.

Hosmer, LT (1995). The connecting link between organisational theory and philosophical ethics. *Academy of Management Review*, 20, 379–403.

Huang, Y and Wilkinson, IF (2013). The dynamics and evolution of trust in business relationships. *Industrial Marketing Management*, 42(2013), 455–465.

Huemer, L (2004). Activating trust: the redefinition of roles and relationships in an international construction project. *International Marketing Review*, 21(2), 187–201.

Jones, GR and George, JM (1998). The experience and evaluation of trust: implications for cooperation and teamwork. *The Academy of Management Review*, 23(3), 531–546.

Kramer, RM (1999). Trust and distrust in organisations: emerging perspectives enduring questions. *Annual Review of Psychology*, 50, 569–598.

Kumaraswamy, MM and Mathews, JD (2000). Improved subcontractor selection employing partnering practices. *Journal of Management in Engineering*, 16(3), 47–57.

Kumaraswamy, MM, Yean, FYL, Rahman, MM and Phng, ST (2005). Constructing relationally integrated teams. *Journal of Construction Engineering and Management*, 131(10), 1076–1084.

Langfield-Smith, K (2008). The relations between transactional characteristics, trust and risk in the start-up phase of a collaborative alliance. *Management Accounting Research*, 19(2008), 344–464.

Larson, E (1997). Partnering on construction projects: a study of the relationship between partnering activities and project success. *IEEE Transactions on Engineering Management*, 44(2), 188–195.

Latham, M (1994). *Constructing the Team*. London: The Stationery Office.

Li, PP (2008). Toward a geocentric framework of trust: an application to organisational trust. *Management and Organisation Review*, 4(3), 413–439.

Liu, AMM, Fellows, R and Nag, J (2004). Surveyors' perspectives on ethics in organisational culture. *Engineering, Construction and Architectural Management*, 11(6), 438–449.

Mayer, RC, Davis, JH and Schoorman, FD (1995). An integrative model of organisational trust. *Academy of Management Review*, 20: 709–734.

Miller, C, Packham, G and Thomas, B (2002). Harmonization between main contractors and subcontractors: a prerequisite for lean construction. *Journal of Construction Research*, 3(1).

Morgan, S (2009). The right kind of bribe. *Building Magazine*, 9 October, pp. 8–9.

Morrell, S (2011). Major strategic priorities and opportunities for construction. *Build Offsite*. Available at: http://bit.ly/120dKeL [Accessed 14th January 2014].

O'Hara, CA (2006). A cognitive theory of trust. *Washington University Law Review*, (84): 1717.

Olomolaiye, PO and Price, ADF (1989). A review of construction operative motivation. *Building and Environment*, 24(3), 279–287.

Oxford Current English Dictionary (1990). Oxford: Oxford University Press.

Patel, H, Pettitt, M and Wilson, JR (2012). Factors of collaborative working: a framework for a collaboration model. *Applied Ergonomics*, 43: 1–26.

Pinto, JK, Slevin, DP and English, B (2009). Trust in projects: an empirical assessment of owner/contractor relationships. *International Journal of Project Management*, 27(6), 638–648.

Rahman, HA, Bari, S, Karim, A and Danuri, MSM (2007). Does professional ethic affects construction quality? *Quantity Surveying International Conference*, 4–5 September 2007, Kuala Lumpur, Malaysia.

RICS (2010). *Maintaining Professional and Ethical Standards*. London: Royal Institution of Chartered Surveyors Publications.

RICS (2012). *Contracts in Use: A Survey of Building Contracts in Use During 2010*. London: Royal Institution of Chartered Surveyors Publications.

Rose, T and Manley, K (2011). Motivation toward financial incentive goals on construction projects. *Journal of Business Research*, 64, 765–773.

Ross, A (2011). Supply chain management in an uncertain economic climate: a UK perspective. *Construction Innovation*, 11(1), 5–13.

Rotter, J (1980). Interpersonal trust, trustworthiness and gullibility. *American Psychologist*, 35, 1–7.

Rowlinson, S and Cheung, FYK (2004). A review of the concepts and definitions of the various forms of relational contracting. In Kalindindi, SN and Varghese, K (Eds), *Proceedings of the International Symposium of CIB W92 on Procurement Systems*, 7–12 January 2004, Chennai, India. 227–236.

Silva, SC, Bradley, F and Sousa, CMP (2012). Empirical test of the trust performance link in an international alliances context. *International Business Review*, 21(2012), 293–306.

Sitkin, RB and Roth, NL (1993). Exploring the limited effectiveness of legislative remedies for trust/distrust. *Organisation Science*, 4, 367–392.

Steers, RM, Porter, LW and Bigley, GA (1996). *Motivation and Leadership at Work*. 6th ed. Singapore: The Graw-Hill Companies Inc.

Tabassi, AA and Bakar, A (2009). Training, motivation, and performance: the case of human resource management in construction projects in Mashhad, Iran. *International Journal of Project Management*, 27, 471–480.

Tabassi, A, Ramli, M, Hassan A and Bakar, A (2011). Effects of training and motivation practices on teamwork improvement and improvement and task efficiency: the case of construction firms. *International Journal of Project Management*, 30, 213–224.

Thorgren, S, Wincent, J and Eriksson, J (2011). Too small or too large to trust your partners in multi-partner alliances? The role of effort in initiating generalized exchanges. *Scandinavian Journal of Management*, 27(2011), 99–112.

Thurairajah, N, Haigh, R and Amaratunga, RDG (2006). Cultural transformation in construction partnering projects. *Proceedings of the Annual Research Conference of the Royal Institution of Chartered Surveyors COBRA*. University College London, 7–8 September 2006. 5–8.

Vaaland, TI (2004). Improving project collaboration: start with the conflicts. *International Journal of Project Management*, 22(2004), 447–454.

Vee, C and Skitmore, M (2003). Professional ethics in the construction industry. *Engineering, Construction and Architectural Management*, 10(3), 117–127.

Walker, A (2009). *Project Management in Construction*. 5th ed. Oxford: Blackwell Publishing Ltd. 150–158.

Walker, D and Hampson, K (2003). Enterprise networks, partnering and alliancing. In Walker, D and Hampson, K (Eds), *Procurement Strategies: A Relationship-based Approach*. UK: Blackwell Science Ltd. Chapter 3. 30–73.

Wong, WK, Cheung, SO, Yiu, TW and Pang, HY (2008). A framework for trust in construction contracting. *International Journal of Project Management*, 26, 821–829.

Wu, S and Udeaja, C (2008). Developing a framework for measuring collaborative working and project performance. In Dainty, A (Ed.), *Procs 24th Annual ARCOM Conference*, 1–3 September 2008. Cardiff, UK: Association of Researchers in Construction Management. 983–992.

8 Critical success factors for construction clients

8.1 Introduction

Projects are known to be complex, often involving a lot of aspects that are carried out through human interactions and entail considerable amount of resources in terms of cost and time. Accordingly, it is very easy for construction projects to perform badly if they are not managed properly and effectively. Unfortunately, it seems that in the vast majority of cases when things go wrong, it is people who are responsible for the poor outcomes. The management of construction projects involves technical elements as well as human elements, with a diversity of individuals from different backgrounds and possessing a wide range of behaviors and intellect. Mismanagement, clashes of personalities, dissimilarities in goals and objectives will no doubt lead and create differences and conflict of interests. If these areas of mismatch are not managed effectively by construction clients, it will be very difficult for project teams to achieve successful outcomes. It is believed that human and soft skills are essential skills, alongside technical abilities, to enable construction clients to achieve successful projects.

In consideration of the above, the aim of this chapter is to investigate the extent to which 'soft skills', also known as human or 'people skills', can influence the performance and success of construction clients. The chapter is predominantly focused around a research study conducted by Baban (2013) which investigated those critical skills required by construction clients, which can have a positive influence on achieving successful project outcomes. In this regard, the study provided important findings to support the following areas of research:

- Determination of the right balance between hard and soft skills for construction clients' success.
- Identification of those key human skills that influence the performance of construction clients.
- Examination of the importance of individual qualities of construction clients to promote project success.

These are the skills and knowledge that construction clients must have and employ to achieve success and realise all project objectives. This chapter is focused on determining the right balance of human skills compared with the technical skills that are perceived to be widely sought after and developed by construction clients. Construction clients can emanate from a number of different backgrounds, depending on the businesses that employ them, but are mostly trained with technical, science based skills at the forefront in their day to day business. Nevertheless, the reality of interacting and dealing with aspects that involve people, human nature and diversity is with universally accepted to be very different from merely possessing technical attributes, at all levels within client organisations. As a consequence, it can be surprising what course of action construction clients take from time to time, which can in some cases end up in heated discussions and adversarial atmospheres.

Understandably, individuals within project teams are usually focused and interested in their respective roles within the construction process. This can sometimes create conflicting situations where roles cross over or different interests or opinions emerge. In such cases, this can result in adversarial situations where individuals try to do their best and 'fight their corners'. It is arguably construction clients, normally acting as 'project sponsors' in their respective leadership roles who should link, coordinate and ensure that project team members act in unity through teamwork principles. They should ensure that the overall contributions and outcomes of project teams are greater than the sum of individual contributions. As such; construction clients are expected to have human and soft skills in addition to construction related technical and hard skills in order to achieve project objectives and successful outcomes. To determine the aim and objectives of this chapter, critical success factors (CSFs) for construction clients have been explored together with the project's success criteria (SC). It is important to differentiate between the criteria that are set at the beginning of projects, for measuring against; and the factors that affect and influence the success of projects. Possessing soft skills together with the correct application of these skills at the appropriate time with the right people may prove to be the difference and become a CSF in managing projects successfully for construction clients.

Human and technical skills are required for construction clients to be able to successfully manage construction projects. This chapter will explain the nature and the scale of the problem within the context of the 'Skill Approach' by Robert Katz.

Critical success factors and success criteria will be discussed in this chapter in terms of what are they, the differences between them, how they affect project teams and how they influence the overall performance of construction clients during the life cycle of projects. However, the main content of the chapter is to explore aspects of human and soft skills within the context of personality, knowledge and skills that construction clients need to have if they were to succeed in achieving the goals and objectives of their projects. Two main areas are explored, which have direct links to project outcomes and construction clients' performance in meeting project and stakeholder objectives. These two areas are the project's SC and CSFs that affect and influence the way that construction clients manage project teams including stakeholders. In addition, areas concerning human and soft skills akin to leadership, teamwork, project management processes and stakeholder engagement are also investigated.

8.2 The articulation of the problem for construction clients

Success is a journey not a destination.
(Ben Sweetland, cited in Goodreads, 2007)

The majority of construction projects have the reputation of failing to meet one or all of the three main aspects of projects' triangle constraints known as the 'iron triangle'. In the United Kingdom (UK), predictability on projects in terms of cost and time have traditionally been poor with approximately half of projects delayed or completed over budget (Constructing Excellence, 2013). Criticism has been directed at the construction industry for its fragmented and disintegrated approach to the way that projects are delivered and the failure to develop and formulate effective team works (Baiden et al., 2006), with project outcomes disappointing stakeholders (Cooke-Davies, 2002).

Construction clients, in a similar way to project managers, are normally drawn from those individuals who possess technical expertise; this facet being commonly perceived as an essential requirement (Hodgson et al., 2011). In some cases, such individuals who have strong technical and business acumen simply lack the soft and human skills necessary to fulfil their roles satisfactorily. Accordingly, Russell (2011) stated that hard skills feature strongly in the training and appointment of construction clients. Despite this, it is considered essential that construction clients should have the necessary human skills to enable them to get along and work effectively with others (Ling et al., 2000). In addition, to allow them to interact with others, Trompenaars and Hampden-Turner (1997) advised that understanding values and beliefs that people hold from different cultures is an important consideration. For these reasons, it is felt necessary that effective and successful construction clients need to have both technical and people skills to perform in their roles. This may highlight an issue that universities and other educational institutions should consider how to embed more human and soft skills studies and subjects into their built environment courses to embed these vital skills at early stages of the professionals' career life. Employers may also consider providing training and support in soft skills as well as hard skills, although this may not be easily implemented perhaps for financial reasons. Lewis (2007) highlighted that this may be the case because there is no immediate return, and benefits cannot be quantified, thus investments and training in soft skills may be regarded as not attractive in the short term.

Delivering successful projects or failing to meet stakeholders' objectives is ultimately the responsibility for leaders including construction clients (Haughey, 2001). There does not appear to be any authoritative definition of the role of construction clients but a close match comes from the project management role. The Association for Project Management (APM) provided a definition as 'at its most fundamental, project management is about people getting things done' (APM, 2012). In addition, according to the APM's Body of Knowledge (BOK), project management was defined as:

> A management discipline that is differentiated from the management of an organisation's business-as-usual by the fact that a project has a clear objective and deliverables, with a defined start and end, that must be completed on time, within budget (cost) and to the agreed quality and, of course, it must deliver the agreed benefits.
>
> (APMBOK, 2006, p. 1)

A different definition by the Chartered Management Institute (CMI) states that 'Project management involves the co-ordination of resources to complete a project within planned time and resource constraints and to meet required standards of quality' (CMI, 2008, p. 1). Unlike the APM and CMI, the Royal Institution of Chartered Surveyors (RICS) pleasantly mentioned the word 'people' in its definition of project management as 'it is the management of people, time and costs by an individual or a team to ensure the efficient commencement, progress and conclusion of a project' (RICS, 2013).

According to Lester (2007, p. 6), it is important for construction clients to have the designated levels of responsibility, authority and accountability assigned to them to facilitate successful project outcomes. Despite this, Lewis (2007) explained that in many cases, construction clients are given a lot of accountability and responsibility but at the same time they lack authority. Accordingly, there is a suggestion that construction clients must be aware and know how to deal with what may be interpreted as a weakness in their positions. For this reason, Walker (2007) explained that the leadership style of construction clients should be democratic, persuasive and influential to gain respect and motivate a potentially large project team.

8.3 Understanding what skills construction clients require

In considering the skills which construction clients require, it is first worth contemplating what the word skills means in practice. Katz (1974) defined skills as 'an ability which can be developed, not necessarily inborn, and which is manifested in performance, not merely in potential'. Skill is also defined as 'the ability to do something well; expertise: difficult work, taking great skill' (Oxford Dictionary, 2013). Furthermore, Virkus (2009) defined skills as 'what leaders can accomplish' and this may present the best definition, given that construction clients are regarded as leaders in the construction procurement process.

The 'Three Skill Approach' theory (Katz, 1974, p. 1) is based on technical, human and conceptual skills and this will be looked at in considering what skills are required by construction clients.

Technical skills are normally easily recognisable, and Katz attributes them to specialisation, expertise and advanced technology, and the majority of training

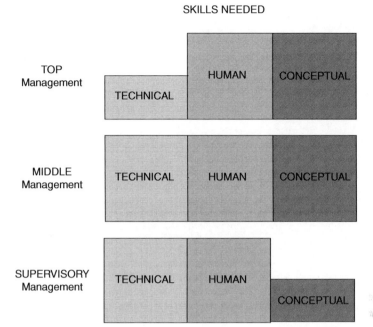

Figure 8.1 Management skills necessary at various levels of an organisation

programmes are focused on developing technical skills. Conversely, human skills involve the ability to work effectively with people as a team member. Conceptual skills involves the ability to see 'the big picture', and includes recognising the dependencies of various functions of organisations and how changes in any one affect all the others.

Katz's theory on how the three types of skills relate and interact concludes that a balance of each type of the three skills is required, and appropriately applied at different levels of organisational hierarchy as shown in Figure 8.1. This relationship should lead to achieving successful outcomes for the organisations, by performing to technical challenges and by understanding and leading people. According to Katz's theory, the more individuals move from operation levels towards management and higher up the scale, the less technical skills are required. Accordingly, it should be noted that despite the importance of technical skills at supervisory and middle management levels, human and conceptual skills are required for those in senior management positions.

8.4 Success criteria on projects for construction clients

The temporary nature of construction normally implies that construction phases of projects are relatively short. In order to measure overall success, projects should

Table 8.1 Content analysis and qualitative themes

No	Theme code for first level observation	Content analysis	Literature source	Group title based on similarities and inter-connections	Observation or proposition of explanation	Data inconsistencies	Data similarities
1	Personal characteristics, respect and charisma	31	Barnes, 1989; Ritz, 1994; Ling et al., 2000; Fisher, 2001; Haughey, 2001; Dvir et al., 2003; APMBOK, 2006; Lewis, 2007; Wong, 2007	Human skills	Technology and industry complexity have undermined human skills and working relationships.	Both sets of skills are equally important and complement each other. And construction clients cannot manage without one or the other.	Construction clients need human skills to get on and work with project teams and bring out the best in people.
2	Key human skills, people management and emotional intelligence skills	53			When practising leadership, construction clients must lead by example.	Having human and management skills may win three quarters of the battle.	Planning and organising is a major and important part of construction clients' job. Successful projects must have plans.
3	Leadership direction and motivation	39			Construction clients need to be accessible and open to people to interact with and look up to.	Hard skills are favoured by technical professionals who progressed to management roles. Hard skills are also favoured by constructors (contractors) rather than consultants.	Construction clients have to be strong, earn respect and trust of project teams and stakeholders to gain power, influence and authority.
4	Planning and organisational skills	14				Construction clients need good people skills rather than technical skills to manage project teams and stakeholders.	The majority of construction clients' role is dealing with people; hence construction clients need to practise management to manage project resources and at the same time to lead, direct and motivate.

5	Macro management and delegation	15			Construction clients have to be competent, have sound knowledge of the industry and have technical skills.
6	Learned capacity and competence	16			Management and hard skills are just as important and essential as human skills.
7	Change and risk management and good judgement	18	Ritz, 1994; Abraham, 2002; Fisher, 2011; Hodgson et al., 2011	Technical skills	The integration of management and leadership is a must for construction clients to deal with the people and the technical elements of the project.	Construction clients need to have technical proficiency to be able to manage technical staff. Construction clients' people skills have more impact than technical skills with project outcomes.	Construction clients must constantly manage change and risk and apply good judgement. Traditional project deliverable aspects (cost, time and quality) are critical but not limited to when success is measured.
8	Micro management and deliverables	24			Projects may exceed the iron triangle limits but still be classed as successful providing other clients' objectives were achieved.		Methodical management of project organisations and resources together with the ability and good judgement to delegate appropriately. Construction clients must have high standards of transferable managerial skills.

(*continued*)

Table 8.1 (Cont.)

No	Theme code for first level observation	Content analysis	Literature source	Group title based on similarities and inter-connections	Observation or proposition of explanation	Data inconsistencies	Data similarities
9	Clients' objective, happy clients, legacy and benefits	38					Achieving clients' objectives including and beyond the iron triangle, getting best value for money and realising the benefits and legacy of projects.
10	Integration and shared understanding	41	De Wit, 1998; Haughey, 2001; Lester, 2007; Walker, 2007; Toor and Ogunlana, 2008; Russell, 2011	Collaboration and common objectives	Success is subjective; one party's success may be another's failure. The priority of one stakeholder may not match all project stakeholders.	Construction clients must achieve the iron triangle and other client and stakeholder benefits. As long as client and stakeholder objectives and benefits are realised and they are happy, achieving the iron triangle may not be important.	Construction clients must integrate project teams by sharing understanding through good communication, teamwork and collaboration. Achieving clients' objectives means happy project environment and repeat business for project teams. Communication leading to shared understanding and common objectives is crucial for success.
11	Teamwork and collaboration	23					Construction clients can't do it alone and need project teams to do the job; teamwork and collaboration is a key factor for success.

12	Natural progression, promotion and gained abilities	25			The majority of construction management professionals come from the technical side of the business. Organisations rely on professional managers who are drawn from technical specialists, who have come up through the ranks. Professionals with strong technical background lack human skills and have difficulties dealing with people. Training and education provide construction clients with skills, knowledge and subsequently confidence to succeed.
13	Education and training	23	Ritz, 1994; Lewis, 2007; Muller and Turner, 2010; Hodgson et al., 2011	Environment and external factors	More emphasis has been placed on technical skills training than human skills training by organisations. Having qualifications, technical skills and knowledge can win half of the battle.

98 *The construction industry and clients*

have a set of predetermined criteria that must be well established, agreed and measured from initiation to completion to realise the benefits. The success criteria provides important data to establish the performance of projects against targets, and is a useful benchmark for construction clients. Traditionally the 'iron triangle' of cost, time and quality criteria has represented the primary success measures for projects, but in recent years other changes to these criteria have been introduced.

Along with these changes, it seems that one thing has not changed; and that is the need for good human and soft skills side by side with the hard and technical skills for construction professionals. This is particularly the case for construction clients who manage teams of individuals in addition to projects and processes.

8.5 Research study to identify the success factors for construction clients

The research study undertaken by Baban (2013) provided valuable information of those factors and traits of construction clients which can have a direct and indirect influence on the success of projects. The study prompted and generated information from different sources and angles representing different roles, interests and approaches for professionals working and managing construction projects. The information sought to provide data on actions, behaviours and attitudes of construction clients. This was in an attempt to identify the cause of problems associated with human and soft skills in conjunction with technical and hard skills.

The analytical process for Baban's research consisted of obtaining feedback from survey questionnaires, in addition to developing qualitative themes, derived from key words from responses on construction professionals during interviews. Nineteen qualitative themes were developed in this regard as shown in Table 8.1. These themes were then grouped together under one title based on relations, skills, interactions and links among the component factors of each group.

The study by Baban identified four main areas of hard and soft skills, grouped under one title or theme, in terms of how individuals act, react and behave. These group titles comprised (i) human skills, (ii) technical skills, (iii) collaboration and (iv) common objectives and environment and external factors, and these are shown in Table 8.2. Each of these themes was ranked in order of their importance and/or influence on the performance and the success of construction clients in meeting project and stakeholder expectations. Table 8.2 illustrates the calculation of the percentage ranking based on the number of labels for each theme with 30.8 per cent of the titles represent humaning skills, 30.8 per cent technical skills, 23.1 per cent collaboration and common objectives and 15.3 per cent for environment and external factors. From Table 8.2, the study by Baban provided findings related to the inter-relationship between hard and soft skills, their importance and their influence on the performance and success of construction clients.

Table 8.2 Percentage rate of theme titles in relation to the Skill Approach

ID	Human skills	Ratio	ID	Technical skills	Ratio
1	Personal characteristics, respect and charisma		8	Macro management and delegation	
5	Key human skills, people management and emotional intelligence skills	30.8%	10	Learned capacity and competence	30.8%
7	Leadership direction and motivation		11	Change and risk management and good judgement	
9	Planning and organisational skills		13	Micro management and deliverables	

ID	Collaboration and common objectives	Ratio	ID	Environment and external factors	Ratio
2	Client's objectives, happy client, legacy and benefits		4	Natural progression, promotion and gained abilities	15.3%
3	Integration and shared understanding	23.1%	6	Education and training	
12	Teamwork and collaboration				

8.6 Analysis and reflection of the study on critical success factors for construction clients

In the early stages of the careers of construction professionals, particular attention and great endeavours are made to acquire, enhance and improve technical knowledge and the skills that they require in managing projects. On this basis, it can be deduced that it is the technical skills of these individuals which needs to be improved and learned upon, rather than managerial skills or human skills. This notion may be more applicable to jobs that involve engineers, architects, surveyors and other technical construction roles. Arguably, the role of construction clients is different from these more technically based staff, and involves more managerial skills and business acumen accordingly.

8.6.1 Human skills versus technical skills for construction clients: human skills

Modern client organisational structures have identified the need for the respective roles of construction clients to be more far reaching, when compared to other disciplines and industries. However, the argument for construction clients is whether human skills are as important as technical skills and whether human skills can be recognised as a critical success factor. Evidence from the findings of Baban (2013) suggested that both types of skills are equally essential and important for construction clients. However, the findings indicate that a complementary and balanced degree of both these skills is required for construction managers to perform well. Human skills are regarded as being essential to enable construction clients to interact with project teams and be able to bring out the best in people. This argument reflects the endorsement of Ling et al. (2000) which reinforced the necessity of human skills to get along and work effectively with others. This was also emphasised by Wong (2007) who found that construction clients as leaders need to motivate teams in order to achieve successful project outcomes.

It is important to clarify that the human or the soft skills relate to social skills that the majority of people have, but unfortunately some may lack. In the latter case, the majority of these skills can be acquired through training, mentoring, experience and practice. Such skills like leadership, team building, communication, respect for others and the ability to listen and learn, should therein be a focus for construction clients throughout their careers. Data findings by Baban (2013) also revealed that the majority of construction clients regarded dealing with people as an influencing factor for project success. Another consideration for success revealed by the study is that construction clients have to be strong but not overbearing to encourage people to approach them. This can earn them the respect and trust of project teams and stakeholders to gain power, influence and authority. This finding was supported by Ritz (1994), which concluded that construction clients must gain authoritative respect and cooperation from project teams to perform well.

Notwithstanding the above, there has been disagreement in the past as to the extent and the balance between the two sets of skills. Certain inconsistencies were acknowledged when some of the data from Baban (2013) found that construction clients need good people management and human skills but few technical skills in order to effectively manage project teams and stakeholders. This position was advocated by Katz (1974), in line with his 'Skill Approach' model. This clearly demonstrates the extent to which human skills are required at all levels of organisation structures, whether it is at the operational, management or executive levels. However, not surprisingly, the data obtained from people who have naturally progressed from technical jobs into management roles, without gaining and obtaining management qualifications and trainings, showed that they favoured technical skills to human skills by a considerable margin. Understandably, their justification was that in order for construction clients or site managers to be able and be competent to manage the construction activities on site, it requires significant levels of technical knowledge and skills.

Within the context of the skill approach, there is an argument that technological and industrial complexity has undermined human skills and healthy working relationships amongst members of project teams and organisations. Some of the data obtained highlighted that having human and management skills may greatly benefit construction clients, but they have to be aware that when practising management and leadership roles they must lead by example. Furthermore, they must display openness, with approachable personalities and be regarded as being accessible.

8.6.2 Human skills versus technical skills for construction clients: technical skills

A considerable weight was attached in the research study by Baban (2013) to the importance of technical skills for construction clients. Some of the survey participants referred to these as transferable skills. The findings highlighted that construction clients must be competent, have sound knowledge of the industry and possess strong technical skills. Moreover, management and hard skills are as just important and essential as human skills, which is a view advocated by Fisher (2011) who stated that 'human skills alone are not the magic bullet to realise success'. Findings have also advocated that there is no value whatsoever for construction clients to be technical experts if they do not have the skill and the ability to work, manage, lead and communicate with their project teams. However, it should be noted that Fisher (2011) simultaneously argued that management of people has become a vital role for construction clients.

8.6.3 Collaboration and common objectives

The 'iron triangle' of cost, time and quality elements often appears to be at the top of the list of priorities and objectives for clients and stakeholders from the outset. Notwithstanding this premise, there is an important role for construction clients in keeping project teams and working environments positive and collaborative, to avoid adversarial situations arising. This may only be achieved by collaborative team working, integration, shared understanding and effective communication.

8.6.4 Environment and external factors

Construction clients should perform their roles and interact with their respective project teams within well defined environments which may have been established by other stakeholders in some cases. A dilemma for construction clients sometimes relates to them lacking human skills within these environments; the majority having emerged from the technical side of their businesses. Some have been promoted within their organisations and risen up the ranks with no management qualifications, training or experience. Consequently, these professionals sometimes lack human skills and have difficulties dealing with people, which is a view articulated by Lewis (2007).

8.6.5 Determining the right balance between hard and soft skills for the success of construction clients

Findings from Baban (2013) identified one common principle and this related to both human and technical skills being important and essential for construction clients. However, different views and judgements emerged on the split and balance between the two sets of skills. The majority of the interviewees from Baban's research study did score these two different skills quite closely in terms of importance. Although there was a minority who favoured hard and technical skills by a considerable margin, the majority of interviewees' judgements were ranging from 40 per cent to 60 per cent in favour of human skills. This assessment is consistent with the literature review findings and observations in Table 8.2. The literature review findings emphasised and insisted on the importance and the major part that human skills can play in the roles of construction clients. Notwithstanding this, it should not be underestimated the extent to which hard and technical skills play an important role in allowing construction clients to succeed in their roles. Accordingly, it is unlikely that construction clients can succeed in achieving successful project outcomes without a reasonable degree of human or technical skills. There was a broad range of agreement amongst the survey participants that technical skills are needed more at operational and middle management levels of organisational structures.

The chapter findings indicated that technology and industrial complexity are contributors in undermining human relationships and skills. There is no doubt that technologies such as the internet, email and other digital technologies have reduced the amount of real physical human interaction when it comes to communication. Technology has arguably led to the absence and lack of human emotions, and where certain situations may require emotions and body language to sensitively or passionately articulate messages more humanly rather than mechanically. Barnes (1989) argued that decades and centuries ago, few technical and hard skills might have been used and applied in the absence of advanced science and technology compared to the level of soft and human skills that engineers and builders had to deal with in managing their projects. This may represent a focus for reflection for construction clients in line with improving their interpersonal skill sets.

8.7 The key human skills that influence the performance of construction clients

The process of analysing the research data and subsequently obtaining the results and findings revealed a number of essential and critical talents, abilities and capacity, which can all be gathered under the human skills umbrella. This objective of Baban's research was designed to identify the key human skills that are most influential and effective in allowing construction clients to perform their roles successfully. One of the key factors which the data uncovered was that construction clients need to have people skills and people management. Elia (2004) argued that it should not be surprising that construction clients' people skills can have more impact than technical skills when it comes to project outcomes. These skills could involve construction clients using charisma and resilience to face challenges, inspire and earn the respect of others. Furthermore, construction clients need to have the ability to interact and relate well with project teams and be open, accessible and approachable to people. This will enable them to more effectively deal with project challenges and problems earlier rather than later when matters may have escalated and potentially spiralled out of control. These statements were supported by Ritz (1994) who found that construction clients must gain authoritative respect and cooperation from people in order to succeed.

Leadership was another critical skill that was highlighted to be one of the most important and key skills for construction clients. Construction clients cannot perform their roles in isolation as they are heavily reliant on the remainder of project teams in all aspects of their work. They need to develop those personable skills for earning trust, providing direction, communicating effectively and motivating project teams. In addition, construction clients need to share their aims, objectives and visions for projects clearly to their teams through clear, effective and timely communication. It is crucial for them to rely on others' expertise to manage projects effectively. For this reason, Wong (2007) argued that construction clients need to identify the potential of their team members and to bring out the very best in people. This aspect was reinforced by Haughey (2001) who stressed that teams must be motivated to do well and to go 'the extra mile'.

The aforementioned key human skills linked to leadership, could not overshadow the importance of management and specifically planning and organisational skills. Baban's research clearly articulated that managing projects requires planning, logistics knowledge and organisational competencies for construction clients. Accordingly, it is imperative that construction clients have strong organisational skills and robust project plans. Dvir et al. (2003) argued that the success of construction clients is heavily reliant on having robust project plans that are closely monitored for successful implementation. This view was supported by the APM (2012) in its BOK publication which underlines the importance of planning and project plans.

8.8 The importance of individual qualities of construction clients for project success

The depth of human and social dimensions of construction clients' personalities, was considered in Baban (2013) to be more important than their general skills. A qualitative approach in the study was able to capture data from face to face interviews where participants were able to share their passions and emotions. They were able to demonstrate clearly and succinctly the importance of individual qualities for construction clients based on their personal experiences and their career journeys to date. Most of the interviewees expressed the importance of care, empathy and understanding of people's emotions, personal circumstances and individual differences. These qualities are particularly important for construction clients as bad experiences for project teams may leave unpleasant memories and legacies as a result of mainly human actions and outputs that could have been avoided. There are views that more pleasant, enjoyable project environments are very important in reputational, economic and financial terms, leading to positive cultures developing on projects.

Construction clients need to be aware and have understanding of personal issues affecting their teams which should enable them to deal and manage individuals more appropriately, effectively and professionally. This argument is supported by Geoghegan and Dulewics (2008) who believed that skills are built on personality characteristics and individual qualities. These personal qualities can be valued and highly respected by team members in terms of appreciation, understanding and respect for different cultural beliefs and religious values. In addition, the survey data found that a number of the practitioners repeatedly referred to the important point of being open and honest with project team members. Fisher (2011) reinforced this view when referring to the importance of ethical qualities for construction clients. Furthermore, the findings are in accordance with the skill approach theory by Robert Katz's 'Skill Approach' model (Katz, 1974).

Despite projects being differentiated on nature and value, they invariably require the skills of effective leadership, the traits model and the Skill Approach model, and Katz's skill model reinforces the concept of both human and technical skills as being essential and required at all levels. Katz's theory has a third skill attached to it, and relevant to construction clients, which is labelled 'conceptual skills'. Notwithstanding this premise, this skill is deemed more applicable to executives and top level management rather than construction clients who are middle managers.

8.9 Summary

Construction clients require different qualities especially focused around social skills, management and interpersonal skills. Management of projects from the perspectives of construction clients has become sophisticated in recent years with development of skills linked to achieving projects' objectives. There appears to be one particular skill that is common and applied to all and this relates to human skills. Arguably, possessing the right degree of human skills allows construction clients to manage any type of projects, and competently carry out many different tasks more competently. The challenge is to understand that the two different types of skills, namely technical and human, are both important and required for successful outcomes. More importantly, construction clients need to strike the right balance between the skills where and when needed, especially given that no project or project environment is similar or the same. Accordingly, success may not be achieved if the right balance of these skills is not present. For this reason, soft and hard skills should be appropriately applied by construction clients at the right time and place on projects. Furthermore, some skills, talents and traits may not always be second nature for construction clients, and for this reason education and training to develop those essential skills may be pertinent.

References

Abraham, GL (2002). *Identification of Critical Success Factors for Construction Organization in the Architectural/Engineering/Construction (AEC) Industry*. Georgia: ProQuest.
APM (2012). *Project Management*. Available at: www.apm.org.uk [Accessed 19th January 2013].
APMBOK (2006). *Planning the Strategy*. 5th ed. Buckinghamshire: APM.
Baban, H (2013). *Critical success factors for construction management*. MSc Construction Management dissertation. University of Bolton, UK.
Baiden, BK, Price, ADF and Dainty, ARJ (2006). The extent of team integration within construction projects. *International Journal of Project Management*, 21, 13–23.
Barnes, M (1989). Fast building: 3 project management. *Architect's Journal*, 71–72.
CMI (2008). *Leading from the Middle*. Corby: Chartered Management Institute: Checklist 041.
Constructing Excellence (2013). *Constructing the Team* (The Latham Report). Available at: constructingexcellence.org.uk/resources/constructing-the-team-the-latham-report/ [Accessed 18th March 2015].
Cooke-Davies, T (2002). The 'real' success factors on projects. *International Journal of Project Management*, 20, 185–190.
De Wit, A (1988). Measurement of project success. *International Journal of Project Management*, 6, 164–170.
Dvir, D, Raz, T and Shenhar, AJ (2003). An empirical analysis of the relationship between project planning and project success. *International Journal of Project Management*, 21(2), 89–95.
Elia, D (2004). *Back to the Basics: Essentials for Today's Construction Project Manager*. Lincoln, NE: iUniverse, Inc.
Fisher, E (2011). What practitioners consider to be the skills and behaviours of an effective people project manager. *International Journal of Project Management*, 29, 994–1002.
Geoghegan, L and Dulewics, V (2008). Do project managers' leadership competencies contribute to project success?. *Project Management Journal*, 39(4), 58–67.
Goodreads (2007). *Find Quotes*. Available at: www.goodreads.com/quotes/search?utf8=%E2%9C%93&q=Success+is+a+journey+not+a+destination&commit=Search [Accessed 16th August 2012].
Haughey, D (2001). *Eight Key Factors to Ensuring Project Success*. Available at: www.projectsmart.com/articles/eight-key-factors-to-ensuring-project-success.php [Accessed 20th December 2011].
Hodgson, D, Paton, S and Cicmil, S (2011). Great expectations and hard times: the paradoxical experience of the engineer as project manager. *International Journal of Project Management*, 29, 374–382.
Katz, RL (1974). Skills of an effective administrator. *Harvard Business Review*, September.
Lester, A (2007). *Project Management Planning and Control*. 5th ed. Oxford: Butterworth-Heinemann.
Lewis, JP (2007). *Fundamentals of Project Management*. 3rd ed. New York: AMACOM, American Management Association.

Ling, FY-Y, Ofori, G and Low, SP (2000). Importance of design consultants' soft skills in design-build projects. *Engineering, Construction and Architectural Management*, 7(4), 389–398.

Muller, R and Turner, R (2010). Leadership competency profiles of successful project managers. *International Journal of Project Management*, 28(5), 437–448.

Oxford Dictionary (2013). *Oxford Dictionaries*. Available at: oxforddictionaries.com/definition/english/skill?q=skill [Accessed 14th December 2012].

RICS (2013). *Project Management*. Available at: www.rics.org/uk/footer/glossary/project-management/ [Accessed 5th January 2013].

Ritz, GJ (1994). *Total Construction Project Management*. Boston: McGraw-Hill.

Russell, D (2011). *Succeeding in the Project Management Jungle: How to Manage the People Side of Projects*. New York: AMACOM, American Management Association.

Toor, S-u-R. and Ogunlana, SO (2008). Critical COMs of success in large-scale construction projects: evidence from Thailand construction industry. *International Journal of Project Management*, 26, 420–430.

Trompenaars, F and Hampden-Turner, C (1997). *Riding the Waves of Culture: Understanding Cultural Diversity in Business*. London: Nicolas Brealey Publishing.

Virkus, S (2009). *Leadership Models*. Available at: www.tlu.ee/~sirvir/Leadership/Leadership%20Models/skills_approach_robert_katz.html [Accessed 10th April 2013].

Walker, A (2007). *Project Management in Construction*. Oxford: Blackwell.

Wong, Z (2007). *Human Factors in Project Management*. San Francisco: Jossey-Bass.

9 The relevance of professional ethics for construction clients

9.1 Introduction

> Consider ethics, as well as religion, as supplements to law in the government of man.
>
> (Thomas Jefferson)

This powerful and unequivocal quotation from the former President of the USA reinforces the importance of ethics in the society and world we live in. Despite the fact that few would disagree with the importance, values and principles of ethics, the practice of professional ethics has however traditionally not been an area for widespread reflection, consideration and focus within the construction industry. Against this background, this chapter of the book will provide further insight into the subject of professional ethics relevant to construction clients and discuss the key areas of ethical standards, values and behaviours, environmental ethics, cultural change, governance and regulation. It will articulate, discuss and analyse some of the problems relating to adopting professional ethics and provide possible reasons to explain and understand why unethical practices exist. The potential importance of ethical principles and potential improvement measures to enhance and improve current practice, especially in the context of the UK construction industry, will also be discussed. Accordingly in this regard the aim and objectives will be to expand the knowledge and breadth of understanding of professional ethics in the construction industry. Finally, conclusions of the study will be presented together with possible recommendations for improvement measures and further research.

9.2 What are professional ethics?

To address the issues of professional ethics, particularly applied to construction clients, one needs to first understand what ethics are and what constitutes ethical or non-ethical practices. Theories of ethics come from a philosopher's perspective and can be categorised as: metaethics, relating to where ethical values and principles emerge from; normative ethics, relating to moral standards of conduct; and applied ethics, involving examining controversial issues (Internet Encyclopedia of Philosophy, 2010).

Ethics has been described in general usage as:

> 'the philosophy of human conduct with an emphasis on moral questions of right and wrong' (Helgado, 2008), 'the system of moral values by which the rights and wrongs of behaviour are judged' (Rosenthal and Rosnow, 1991) and 'a moral philosophy that involves systematizing, defending and recommending concepts of right and wrong behaviour'.
>
> (Internet Encyclopaedia of Philosophy, 2010)

When considering '*professional*' ethics specifically, Lere (2003) describes it as 'a way and approach between professionals and experts as experts and clients as lay people'. Conversely, when we examine professional ethics specifically within a construction context the RICS defines ethics as 'a set of moral principles extending beyond a formal code of conduct' (RICS, 2010). In terms of how it should be applied in the workplace the code of ethics for project managers stresses that 'it is vital that Project Managers conduct their work in an ethical manner' (Walker, 2009). Clearly this highlights that there are many different definitions and meaning of ethics and the possible problems of ambiguity and meaning. This view was supported by Vitell and Festervand (1987) who advocated that complications arise and ethical dilemmas exist as there is no universally accepted definition of ethics. Perhaps this is one of the problems when considering professional ethics in the construction industry on a global scale which could be giving rise to different interpretations of ethics for construction clients and what they mean in practice. This is arguably of particular importance when considering the construction industry, as the understanding and views of many different professional bodies, association and organisations and their members are not always the same. The RICS Professional Ethics Working Party (RICS, 2003) accepts that this potential ambiguity and inconsistency could create problems for the profession. Perhaps therefore a common framework for managing ethical dilemmas and thereby improving ethical standards is required which could address this problem. This opinion is supported by Liu *et al.* (2004) who argued that professional ethics are based on the subjective nature of principles, standards and values which vary between different sectors of the industry and a more consistent approach is required accordingly.

9.3 The importance of professional ethics for construction clients

In order to discuss professional ethics in context it is important to understand why it is important for construction clients. There is a view that ethics are a vital and essential practice requirement as it engenders the general public's trust and preserves their employers' interests (Rahman *et al.*, 2007). Construction clients who act in an ethical manner will enhance their performance which will increase the success of projects. Also for construction professionals to survive it will demand public confidence which is dependent on the ethical conduct and professional knowledge (RICS, 2010). Accordingly this could suggest that any compromises on ethics could jeopardise the service delivery and damage public perception and the image of the construction professionals.

One of the other problems the construction industry is faced with lies in its reputation and the general perception of the public (Robson, 2000). This has largely emanated from negative press coverage, particularly in the UK, for unethical practices and behaviour uncovered in the past. From data published by the Building Research and Information Service it would seem that contractors in comparison with other construction professionals have gained a greater reputation for unethical behaviour in the past with a higher number of legal disputes emanating from their behaviour and practices (Vee and Skitmore, 2003, p. 120). Research studies conducted by Liu et al. (2004) have suggested that developers and contractors display little emphasis on the existence of ethical codes when compared to other construction professionals such as architects and surveyors. This may therefore suggest that contractor led organisations such as the Chartered Institute of Building (CIOB) should develop and promote cultural change to improve standards and raise awareness of ethical standards. Irrespective of whether the contractors are at fault, reputational damage to construction clients may ensue when controversy on projects, culminating from their contractors' wrong doing, emerges in the press. There have been many examples where clients have been held to account for the practices carried out by their appointed contractors. The example of Carillion's demise in 2018 in the UK, led to speculation that public sector organisations, who awarded framework contracts for schools and hospitals, had been negligent in not undertaking more robust financial checks in their due diligence procedures.

A possibly more controversial explanation for unethical practices in contracting could be the relatively high turnover of construction company personnel and the perceived motivation of greed for increased profits (Fellows et al., 2004). Construction companies are frequently accused of only being concerned with short term economics or simply blinkered to the interests and wellbeing of their clients and associates in the industry. After all, why would construction companies become unduly concerned for their clients and practices when in most cases they move from one job to another in short succession? Arguably this is a misunderstood area in construction management. It can produce adversarial management styles geared to aggression and deception rather than a professional approach underpinned by integrity, honesty, transparency, fairness and trust (Walker, 2009). Accordingly this may explain why the importance and practice of professional ethics does not often feature very highly on the agenda of some building contractors. Construction clients should be aware of such negative aspects and manage their selection processes to appoint contractors who have demonstrated social and moral responsibility in the past.

One recent example of controversy is related to construction workers being added to a database and blacklisted by up to 40 major national construction companies. It was reported that this could have excluded many individuals and companies from employment without their knowledge. This followed previous failures of standards and ethics related to The Office of Fair Trading's investigation and report of September 2007 which found that 103 construction firms had colluded with competitors in bid rigging to secure construction contracts. This bid rigging practice is further discussed by Vee and Skitmore (2003, p. 117) who explained that tendering has traditionally been the prime area of concern for unethical practices in the construction industry.

The aforementioned cases involving main contractors does not infer by any means that construction clients always comply with ethical codes. There have been cases in the past which involve construction clients, using their powers as 'paymasters' on projects, to pressure their professional consultants and main contractors to accept fixed price commissions and tenders which do not reflect reasonable margins for them to survive. Such win-lose scenarios, devoid of partnering/teamwork philosophies, could lead to reputational consequences for construction clients under such scenarios. Clearly this is damaging to the construction industry at large in terms of public confidence and trust in the sector. While speculation exists as to the existing law on such matters these cases certainly highlight the debate on whether ethical standards and codes of conduct within the construction industry are being maintained.

9.4 Codes of ethics for construction clients

Owing to increasing concerns in many high profile cases including those previously referred to in this chapter, demonstrating dishonesty and corruption, it is important for construction clients to commit to and encourage project teams to comply with sustainable ethical principles. Codes of ethics which have been introduced have provided an indicator that organisations and institutions take ethical principles seriously as they should outline expectations for all personnel with regard to ethical behaviour and intolerance of unethical practices (Chartered Management Institute (CMI), 2013).

Relationships between construction clients and the professional consultants and contractors they appoint rely on professional ethics and trust especially since fee agreements cannot accurately specify all financial contract contingencies for possible additional services (Walker, 2009, p. 157). The main motivation for the public relying on members of professional bodies relates to them giving advice and practising in an ethical manner (RICS, 2010). Accordingly, the RICS has developed 12 ethical principles to assist their members in maintaining professionalism and these relate to honesty, openness, transparency, accountability, objectivity, setting a good example, acting within one's own limitations and having the courage to make a stand. In order to maintain the integrity of the profession members are expected to have full commitment to these values. Furthermore, a 'Code of Ethics Checklist' published by the CMI sets down that ethics are particularly relevant to maintaining the reputation of an organisation and inspiring public confidence in it (CMI, 2013). For this reason, codes of ethics should reflect the practices and cultures which construction clients want to encourage for their respective organisations and project teams. This is supported by the CMI who advocated that: 'A code of ethics is a statement of core values of an organisation

Table 9.1 Top 11 most frequent unethical practices

Ranking of most frequent acts of unethical conduct	Ranking of frequency
Underbidding, bid shopping, bid cutting	1
Bribery, corruption	2
Negligence	3
Front loading, claims game	4
Payment game	5
Unfair and dishonest conduct, fraud	6
Collusion	7
Conflict of interest	8
Change order game	9
Cover pricing, withdrawal of tender	10
Compensation of tendering cost	11

Rank no. 1 = Most frequent
Rank no.11 = Least frequent
Source: Adapted from Rahman et al. (2007).

and of the principles which guide the conduct and behaviour of the organisation and its employees in all their business activities' (CMI, 2013).

Arguably the main deficiencies of codes of ethics had emanated from the notion that there are no universal standards and accordingly they vary between countries and different sectors in the building industry. Boundaries and barriers created by fragmentation and differentiation within the construction sector have possibly deterred any common frameworks of professional ethics emerging in the past (Walker, 2009). This is an area that demands more attention through multinational dialogue across all areas of the construction sector. One attempt to address unethical behaviour in this way comes from The Global Infrastructure Act Anti-Corruption Centre, which has published a guide with examples of corruption in the infrastructure sector to assist practitioners. It sets down potentially criminal acts of fraud which include collusion, deception, bribes, cartels, extortion or similar offences at pre-qualification and tender, project execution and dispute resolution (Stansbury, 2008). Furthermore, Rahman et al. (2007), from studies conducted on construction professionals, published rankings for the top 11 most frequent unethical practices. These are contained in Table 9.1 and include underbidding, bribery and collusion.

9.5 How should construction clients recognise unethical practices?

The next issue and potential problem concerns the problem of what constitutes unethical behaviour in practice. As previously highlighted there is no universal theory of ethics with different cultures existing within the construction industry and this creates problems and dilemmas in what is ethical or non-ethical (Liu et al., 2004). Clearly this reinforces the need for construction clients to have a consistent approach to professional ethics which can be applied across the whole industry. Liu et al. (2004) explained, however, that this notion of achieving consistency is linked to the different cultures which exist within the construction industry and this may make boundaries between ethical and non-ethical behaviour become blurred at times. A practical example of this could include the boundary between receiving a seasonal gift as a polite gesture and what is deemed to constitute an act of bribery to influence an award of a contract, for instance. Vee and Skitmore (2003, p. 119) attempted to address this potential grey area and offered clarity on the boundary between gifts and bribery. They concluded that gift giving transfers become an illegal act of bribery when they compromise relationships between the gift giver and receiver and favour the interests of the gift giver. This is an important aspect for construction clients, especially at tender stages when bidders may offer them gifts or invitations to corporate functions, to gain competitive advantage over their competitors. It is normal for construction clients to have to sign up to anti-bribery legislation and declare gifts to avoid accusations of impropriety in such cases. Other forms of unethical behaviour could include breach of confidence, conflict of interest, fraudulent practices, deceit and trickery and may also in some cases have problems of grey areas and interpretation difficulties. Moreover, less obvious forms of unethical behaviour could include presenting unrealistic promises, exaggerating expertise, concealing design and construction errors or overcharging (Vee and Skitmore, 2003, p. 120).

9.6 The need for construction clients to uphold cultural values when procuring projects

The next important factor that arguably has a strong influence and link to professional ethics is culture and the cultural values of the construction industry and those construction client organisations that work within it. Adversarial attitudes in the construction industry have affected relationships, behaviours, culture and trust. A major contributor in improving cultures within the construction industry has been professional ethics which defines rules of conduct. However, construction clients should regard their scope and ethical responsibilities as much greater and more extensive than just simply concerning conduct, institutional rules and regulations (Walker, 2009, p. 145).

There have been opposing views on how best construction clients can instigate cultural change within the industry and differentiation and fragmentation can again pose difficulties in this regard in aligning cultures, beliefs and standards (Lui et al., 2004). A practical example of this could be subcontractors having completely different standards, beliefs and values to main contractors and similar scenarios existing between surveyors and architects. This raises the issue of the importance of changing the culture of the industry as a whole and the way it works. If cultural change is required then a further question is raised as to how do construction clients working with their project teams achieve this? The answer could be in improved training, education and personal development to raise the awareness of the importance of professional ethics. Ahrens (2004) certainly supported this view and advocated the use of modules designed to teach ethics to built environment undergraduates to expand their knowledge and understanding of ethical issues affecting the construction industry, with particular emphasis on contracting responsibilities and liabilities. He explained that too many young practitioners graduating from higher education do not possess the skills in areas relating to ethical values, moral working, cultural difference and environmental responsibility and is facilitating modules across 15 European Universities to attempt to address this educational imbalance. Lui et al. (2004, p. 446) presented a similar argument to improve ethical self-regulation and cultural change through education and training. Further supporting views emanated from Vee and Skitmore (2003, p. 117), who explained that ethical codes alone are not sufficient to maintain ethical conduct. Their findings indicated that employer led training and institutional CPD to educate members on what ethical codes mean from a practical perspective can greatly increase awareness and participation in ethical practice.

9.7 Governance and regulation of professional ethics

The next important area for consideration by construction clients is governance and regulation of professional ethics. Behaviour of construction professionals is controlled and monitored by a strict code of ethics, created and maintained by their professional institution (RICS, 2003). These codes of conduct are primarily to ensure that integrity and ethics are maintained at all times. As referred to earlier in this chapter, institutions such as the RICS and RIBA have royal charters which strictly set down rules and regulations relating to professional standards, moral and ethics which all members must comply with. Table 9.2 lists the RICS codes of conduct and practical measures to ensure compliance in this regard. These include such elements as conflict of interest, corruption, confidentiality, honesty and integrity. Members of professional bodies are bound by codes of conduct to address the issue of non-ethical behaviour and to attempt to provide a context of governance (Liu et al., 2004). Members of such institutions are then bound by such codes of conduct in the way they practise and the institutions reserve the rights to take action against members who breach rules and regulations laid down. To put this into context the RICS Professional Regulation and Consumer Protection Department has in the past dealt with approximately 2700 cases of professional misconduct mostly related to breaches in regulations, conflicts of interest and accounts breaches (RICS, 2010).

It is important for construction clients to be aware that adopting the aforementioned codes of conduct gains and maintains respectability and integrity for project teams, and their respective professional institutions and organisations (Vee and Skitmore, 2003, p. 117). Stewart (1995), however, was sceptical of such institutional codes of conduct and explained that these are merely guidelines for professionals to interpret as they wish and do not promote values, ethics and morality accordingly. Clearly, corrupt behaviour is subject to more than just policing by professional institutions and in some cases can be deemed criminal offences. This being the case, there is a strong argument to suggest that the law, coupled with institutional sanctions, may present an acceptable level of deterrent in that professionals will think twice before committing unethical behaviour if consequences are considered grave enough. Lui et al. (2004) argued, however,

Table 9.2 RICS codes of conduct

RICS codes of conduct	Measures to ensure compliance
Act honourably	Never put your own gain above the welfare of your clients or others to whom you have a professional responsibility. Always consider the wider interests of society in your judgements.
Act with integrity	Be trustworthy in all that you do – never deliberately mislead, whether by withholding or distorting information.
Be open and transparent in your dealings	Share the full facts with your clients, making things as plain and intelligible as possible.
Be accountable for all your actions	Take full responsibility for your actions and don't blame others if things go wrong.
Know and act within your limitations	Be aware of the limits of your competence and don't be tempted to work beyond these. Never commit to more than you can deliver.
Be objective at all times	Give clear and appropriate advice. Never let sentiments or your own interests cloud your judgement.
Always treat others with respect	Never discriminate against others.
Set a good example	Remember that both your public and private behaviour could affect your own, RICS' and other members' reputations.
Have the courage to make a stand	Be prepared to act if you suspect a risk to safety or malpractice of any sort.
Comply with relevant laws and regulations	Avoid any action, illegal or litigious, that may bring the profession into disrepute.
Avoid conflicts of interest	Declare any potential conflicts of interest, personal or professional, to all relevant parties.
Respect confidentiality	Maintain the confidentiality of your clients' affairs. Never divulge information to others unless it is necessary.

that transgression will still prevail if detection of breaches is considered unlikely or disciplinary measures imposed for breaches are regarded as insufficient or too lenient. Conversely, there is an argument that ethical codes of conduct should not be regarded negatively as a framework for punishing breaches but positively in assisting professionals in recognising their own moral parameters (Henry, 1995). Accordingly, this raises the question of the role of codes of conduct and whether their purpose should be more closely linked to promoting compliance. This could involve motivating professionals to behave in an ethical manner as opposed to them being seen as frameworks to impose punishments for potential breaches.

The relevance of professional ethics 115

9.8 Environmental ethics

Construction clients in their respective leadership roles, when procuring projects, should be aware of the growing importance of environmental ethics, as part of their organisations' social responsibilities. In considering what constitutes environmental ethics one definition is that: 'Environmental ethics concerns formulation and moral obligations regarding the environment' (Tagawa, 2005, p. 12). Increasing awareness of environmental ethics has in recent years occurred owing to the potential environmental problems including resource depletion and climate change. Environmental ethics reflects the impact of policies and actions on environmental matters and future individuals' wellbeing. There is an argument that the construction industry plays an important part in sustainability, conservation and environmental management and environmental ethics should accordingly be particularly relevant to the ethos of projects (Tagawa, 2005, p. 12). Owing to the above factors, there is a strong argument that construction clients should be increasing political influence in this area, and therein drive the agenda forward. This could take the form of instructing their project teams to adopt environmental and sustainable design measures into their projects as one example. This could lead to changes in practice and have reputational benefits for their organisations. However, if construction clients adopt policies which encourage use of non-sustainable construction strategies, this could have negative connotations in their environmental credentials and lead to reputational damage.

9.9 Summary and conclusions

This chapter has discussed many different aspects and issues that influence and affect professional ethics and the repercussions that can arise from non-ethical practices, especially in the context of the UK construction industry. Construction clients, as leaders in the procurement of projects, should be leading the way in cultural changes to improve the reputation of the industry. In this pursuit, they should be aware of the importance of ethics, alternative definitions and interpretations of ethics, the reputation of the construction industry, codes of conduct and governance and regulations in avoiding bad practices. In conclusion, it would appear that measures to improve the practice of professional ethics, such as professional codes of conduct, have gone some way to improve the way the industry works but there are still far too many cases emerging of unethical practices that are blighting the sector. Although arguably these practices are emerging from a small minority of the sector, they have created and may still be creating a bad press for the whole industry and further measures should be instigated by construction clients to address this dilemma. Traditional responses in the past, at an institutional level, have been based on governance, regulations and punishment for non-compliance and clearly these have had only limited success. Perhaps construction clients should be leading the way for a cultural change in the industry to train, educate and motivate construction individuals and organisations. Such changes could be designed around what professional ethics entail, measures to ensure compliance and the benefits that they can bring for the sector. This could be achieved through more focus on further education and higher education course modules linked to professional ethics and CPD through workshops and training events in the workplace. These measures will hopefully contribute to providing a more ethical environment for the industry to work within. It is accepted, however, that to bring about these cultural changes will take conviction, integrity and in some cases courage not to engage in established unethical practices. These improvements, once ingrained within the industry, could then reap massive rewards in providing a safer, honest, trusting and more enjoyable working environment for all.

References

Ahrens, C (2004). Ethics in the built environment: a challenge for European universities. *ASEE Annual Conference Proceeding.* 5281–5289.

Chartered Management Institute (2013). *Codes of Ethics Checklist.* London: Chartered Management Institute.

Fellows, R, Liu, A and Storey, C (2004). Ethics in construction project briefing. *Science and Engineering Ethics,* 10, 289–301.

Helgado, H (2008). The ethical dimension of project management. *International Journal of Project Management,* 26, 743–748.

Henry, C (1995). Introduction to professional ethics. *Professional Ethics and Organizational Change,* 13.

Internet Encyclopaedia of Philosophy (2010, 24 March). www.iep.utm.edu. [Accessed 21st February 2013, from Ethics and Self-Deception].

Lere, JC (2003). The impact of codes of ethics on decision making: some insights from information economics. *Journal of Business Ethics,* 365–379.

Liu, AMM, Fellows, R, and Nag, J (2004). Surveyors' perspectives on ethics in organisational culture. *Engineering, Construction and Architectural Management,* 11(6), 438–449.

Rahman, HA, Bari, S, Karim, A and Danuri, MSM (2007, 4–5 September). Does professional ethics affect construction quality? *Quantity Surveying International Conference.* Kuala Lumpur, Malaysia.

RICS (2003, 1–2 September). COBRA 2003, *Proceedings of the RICS Foundation Construction and Building Research Conference.* University of Wolverhampton. RICS.

RICS (2010, 1 April). *Maintaining Professional and Ethical Standards.* London: RICS.

Robson, C (2000). Ethics: a design responsibility. *Civil Engineering,* 70(1), 66–67.

Rosenthal, R and Rosnow, RL (1991). *Essentials of Behavioral Research Methods and Data Analysis.* 2nd ed. Boston: McGraw-Hill.

Stansbury, CS (2008). *Examples of Corruption in Infrastructure.* London: Global Infrastructure Anti-Corruption Centre.

Stewart, S (1995). The ethics of values and the value of ethics. *Whose Business Values,* 1–18.

Tagawa, S (2005, 27–29 September). Environmental ethics and project management of architecture. *The 2005 World Sustainable Building Conference.* Tokyo. 2–17.

Vee, C and Skitmore, M (2003). Professional ethics in the construction industry. *Engineering, Construction and Architectural Management,* 10(2), 117–127.

Vitell, C and Festervand, D (1987). Business ethics: conflicts, practices and beliefs of industrial executives. *Journal of Business Ethics,* 6, 111–122.

Walker, A (2009). *Project Management in Construction.* Oxford: Blackwell Publishing Ltd.

10 The influence of construction clients on motivating project teams

10.1 Why should construction clients be concerned about project team motivation?

> Nothing great was ever achieved without motivation and enthusiasm.
> (Ralph Waldo Emerson, 1867)

With reference to the above quotation, Challender (2017) concluded that collaboration, trust and successful outcomes on construction projects are strongly affected by motivation levels of project teams including clients, consultants, main contractors and subcontractors. Accordingly, this should be an important area for construction clients in their respective leadership roles to be acutely aware of and address measures to ensure motivation levels remain high throughout the life of projects. Despite this, Herzberg et al. (1959, p. 67) concluded that: 'Construction contractors in particular are only interested in short-term financial incentives and maintaining morale and motivational levels accordingly will only have limited influence on the success of building projects and construction productivity.'

This is a longstanding, arguably misinformed and possibly controversial view held by some within the construction industry. Such an opinion could represent the root cause of those on the lower tiers of the contracting organisation sometimes being accused of 'only being in it for the money' or even worse being typecast as lazy, having no desire to 'do a job well' and uninterested in anything other than financial recompense. This being the case could possibly explain the reason why the practice of motivational management of project teams has traditionally not been an area for widespread reflection, consideration and focus within the construction industry. Other potential reasons could be that construction contractors are sometimes perceived as only being concerned with short term economics or simply blinkered to the plights and wellbeing of their blue collar employees. After all, why would construction companies become unduly concerned for motivation levels and morale of construction operatives who are at the 'bottom of the hierarchy ladder', employed mostly on temporary contracts and who will in most cases move from one job to another in short succession? Arguably this is a misunderstood area in construction management, which should be focused on by construction clients, in their aspirations to secure more successful project outcomes.

Such negative perceptions can produce management styles geared to aggression rather than persuasion and accordingly the importance of motivation and the practice of motivational management of the workforce do not often feature very highly on their agenda. Also, the construction industry has sometimes considered that motivation is not an issue for construction teams, especially those on the lower tiers. This is based on the misnomer that low rates of industrial action and strikes in the construction section have suggested that workers are motivated when compared to their counterparts in other industries such as manufacturing. However, Olomolaiye and Price (1989, p. 280) argued that this historically low industrial action could be the result of fragmentation of the industry and poor construction operative cohesion rather than workers being motivated.

10.2 The bespoke and unique nature of the construction industry

In considering motivation, we need to understand the context and environment within which building operatives operate. In this regard, the construction industry is unique and more complex than most other sectors such as manufacturing, as projects tend to be bespoke and require creative thinking and innovation at most times. Tabassi *et al.* supported this notion in that: 'The construction industry is considered to be one of the most dynamic and complex environments as it is a project-based industry within which individual projects are usually built to client's needs and specifications' (Tabassi *et al.*, 2011, p. 214).

Tabassi *et al.* (2011) also explain that the construction industry, being a project based environment, has many challenges including unique one off products won at short notice, transient workforce moving between different work locations, being male dominated and with a macho culture/climate. Other challenges for the sector include short term teams forming and disbanding, changing skill and competency requirements and increased client pressures over recent years. Tabassi *et al.* also highlighted that the construction sector is sensitive to organisational changes possibly brought about by changing market conditions. What this symbolises is the extremely complex environment that the construction industry operates within, with ever increasing demands and expectations for successful projects in terms of cost, time and quality. In consideration of these challenges and pressures, can construction clients rationally ignore the motivational issues affecting construction staff? Despite these obvious complexities inherent within the sector, the importance of motivating the workforce has been largely undervalued in the construction industry when compared with other sectors such as commerce and manufacturing. When considering this, it is surely worth construction clients contemplating the benefits that could ensue from having a more motivated workforce involved in construction projects. Accordingly, construction clients should be able to identify and analyse the factors which affect motivation levels of construction staff and potential benefits that motivation can bring to construction management. Such awareness should not be confined to construction managers, but extend to those individuals based on site, as they by far represent the majority of individuals making up the workforce on UK based construction projects.

10.3 Factors which affect motivation levels of project teams

Table 10.1 articulates research finding by Kaming et al. (1998, pp. 133–134) on five different categories of construction staff in Indonesia. It can be determined from this research that there are many high ranking factors which are deemed to motivate construction workers other than simply monetary rewards. These include good relationships, safety programmes and supervision. Undoubtedly, the factors affecting motivation levels of all construction staff are endless and accordingly this chapter will specifically consider, analyse and evaluate four main factors, namely communication, training, culture and financial rewards.

Construction clients need to be aware of alienation and the repetitive work of construction staff and the effects on motivation levels through monotony and boredom. Construction clients could request that their contracting partners address this through job rotation and introducing variable tasks and variety of work initiatives where practicable. In addition, the structure of organisations, communication lines, behaviours and the culture that exists can present problems for the motivation levels of site operatives. There are arguments to suggest that these issues are for contractors to manage. However, low motivation levels within construction team workforces can have a detrimental and negative effect on project outcomes in terms of conflict, productivity, poor quality of construction work and low esteem.

Table 10.1 Rank order of job satisfiers for a range of construction personnel

Job satisfiers	Ranking of importance				
	Owners/ clients	Project managers	Site managers	Foremen	Operatives
Financially lucrative	1st	1st	–	–	–
Satisfied customer	2nd	3rd	–	–	–
Job completed on schedule	3rd	–	–	–	–
Tangible physical structure	4th	5th	–	4th	3rd
Good workmanship	5th	2nd	2nd	3rd	1st
Owner satisfied	–	3rd	–	–	–
Good working relationship	–	–	5th	5th	–
Maintain the job	–	–	1st	–	–
Meeting the challenge	–	–	2nd	–	5th
Job costs below estimate	–	–	4th	–	–
Challenge of running the work	–	–	–	1st	–
Maintain the schedule	–	–	–	2nd	–
Productive day	–	–	–	–	2nd
Social work relation	–	–	–	–	4th

As referred to earlier in this chapter, factors that have an influence on the motivation of construction project teams include communication strategies, management styles, training, behaviours and organisational cultures. Construction clients should be aware of how these individual factors will be managed on projects and the effects that these can have on the motivational levels of construction teams.

In considering motivational factors affecting project teams it is important first to consider what motivation actually means. Tabassi and Bakar (2009, p. 474) outlined that: 'Motivation may be defined as the characteristic of an individual willing to expend effort towards a particular set of behaviours.' When considering the potential benefits that improved motivational levels could reap for the industry Tabassi and Bakar (2009, p. 475) explained that the success of construction projects largely depends upon the quality and morale of the people employed by contracting organisations. This view was supported by Steers *et al.* (1996, p. 67) who outlined that motivation and communication can have a direct effect on the performance of employees. Brooks (2006, p. 66) also explained that employees need to be confident and comfortable within an organisation in order to produce successful results and engage in teamwork.

10.4 Communication as a motivational factor

The first factor that construction clients will consider with regard to motivation of project teams is communication. Steers *et al.* (1996, p. 78) outlined that communication can have a direct effect on motivation and therein the performance of construction employees. This is supported by Carnall (1999, pp. 111–134), who explained that poor communication lines and behaviours where employees feel that they are not being consulted or made aware of decisions which affect their work can severely affect motivation levels. Furthermore, he explained that the same outcome could arise where the employees feel they are not being given the sufficient level of authority to take decisions. This when applied to the construction industry could explain why some construction industry employees feel they are not being fully consulted, or 'left in the dark', leading to feelings of demoralisation. Too many times managers in the construction process fail to keep others informed of such essential matters as timescales, coordination, installation dates and quality which can leave them feeling isolated, frustrated and undervalued. These negative factors can result in poor motivation levels and a downward cycle leading to further demoralisation, lower morale and poor project outcomes.

In addition, construction staff could become resentful and anxious if they do not find out information first hand and feel they are being bypassed or undervalued through not being consulted (Brooks, 2006). This can exist within organisations that employ autocratic styles of management. Conversely, Brooks did, however, concede that low self-motivation levels in such cases do not necessarily lead to poor performance. He stressed that employees could still focus on, and achieve success for other reasons such as fear of reprisal, intimidation or job security rather than the desire to do a good job and job satisfaction gained in the process. This could arguably be classified as survival behaviour and an approach which could be described as satisfying the first tier of Abraham Maslow's hierarchy of needs linked to safety and security. Another argument for playing down the significance of motivation could occur where construction projects are of a minor nature leading to short durations for subcontract works especially. In such cases, applying the ideas of Brooks, one could argue that motivational levels of construction staff are less important than they would ordinarily be on longer projects where relationships and interaction between individuals is paramount to success.

Strands of motivational theory relate to the premise that all staff have desires to be treated the same as their colleagues. Where this does not happen it could result in employees feeling less valued than their co-workers. This in turn could lead to implications of reduced effort and productivity, and in some cases employees leaving their respective organisations altogether. Construction staff are considered by some to be at the bottom end of the hierarchy structure of projects on a par with site labourers. Accordingly, applying Adams theory (as cited in Deci, 1975), this could explain why construction staff feel they are treated differently to more senior colleagues, e.g. site managers and as such not worthy of inclusion in the decision making or consultation processes. Where this occurs construction staff can feel their specialist inputs are not welcomed or important, leading to demoralisation and low motivation levels. In such cases implications can include poor productivity, quality, teamwork and cooperation from construction staff on all aspects of projects. Construction clients, as leaders and normally 'project sponsors', should be aware of such negative aspects and manage them accordingly.

Turner et al. (2003, as cited in Dwivedula and Bredillet, 2010, p. 160) identified that motivational initiatives linked to communication include giving employees good clarity, feedback on performance and a feeling of achievement and interaction. Applied to construction projects, communication could be improved in this way by including all project teams in regular briefings and meetings to discuss and agree how their important inputs interrelate to the project as a whole. Construction clients' interventions in enabling dialogue in this way can bring a sense of 'ownership' or 'buy in' for projects. In addition, the use of regular appraisals of construction staff with their line managers could enable individuals to know where they stand within projects and therein create a sense of teamwork and camaraderie. Furthermore, giving construction employees acknowledgement and praise for good work undertaken could prove a highly important factor leading to an upward cycle of performance and motivation. However, this communication and recognition for good performance, unfortunately, is something that is seldom offered to construction staff. The possible reasons for this could emanate from construction clients expecting good results without them providing any positive interventions or underestimating the benefits that praise and feedback can bring to the morale and motivation of staff.

Nesan and Holt (1999, as cited in Tabassi and Bakar, 2009, p. 474) placed emphasis on recognition as an effective means for inspiring motivation and enthusiasm amongst employees. They explained that it is particularly successful when applied to teams as opposed to individuals. Considering this perspective, in the context of the construction industry, initiatives could be introduced on projects by construction clients to acknowledge and in some cases reward construction staff or subcontract teams for outstanding work undertaken. An example of this could be a 'subcontractor team of the month award'.

10.5 The benefits of training and education

Construction clients should be aware of the advantages in keeping project teams motivated by encouraging a culture of staff development through training and education. The potential benefits from the training and education of project teams have been much debated over recent years. Alderfer's theory of motivation is based on existence, relatedness and growth (Deci, 1975) and this could be applied to all construction staff. He concluded that individuals have needs to develop and where they feel that this is not being met they regress to satisfying lower needs such as respectability or safety needs. It should therefore be not surprising that research from the Construction Industry Council (2008 survey of 100 top construction companies) found that those construction companies that have invested in training and staff development initiatives for their staff have over time benefitted. In this regard, reports suggest from increased cooperation from them and more willingness to work collaboratively with the other members of the project team. This was supported by Dwivedula and Bredillet (2010, p. 160), who explained that providing staff with training will provide them with opportunities for further learning and growth which will create an upward cycle of motivational growth linked to career progression and further learning aspirations. Furthermore, Vroom's 'Expectancy' theory could be applied to training and staff development in that this 'investment in people' could improve the belief and self-confidence of construction staff in their individual capabilities. Vroom's theory (BusinessMate.org., n.d.) is particularly relevant as, being multidimensional, it can be related to all levels of construction staff.

Construction clients should be aware that one of the factors adversely affecting motivation levels of construction staff is linked to alienation, where operatives are undertaking jobs which could be classified as repetitive, routine, non-challenging or non-mentally stimulating manual work. It is sometimes forgotten that construction workers, like most other individuals, benefit from being interested in their work and where they feel they are developing their skills in other areas. Accordingly, this should be addressed wherever possible by construction clients, by instructing project teams to create opportunities to vary work by giving construction staff different tasks from time to time through job rotation. This can raise motivation levels by workers learning new skills and has proved to reduce levels of boredom and monotony. Dwivedula and Bredillet (2010, p. 163) reinforced this view and identified facets of training which have a beneficial impact on motivating the workforce that could be applied to contractors, including job advancement, variety and level of knowledge, participative decision making, developing competencies and increased sense of achievement. One could, therefore, surmise that construction clients should encourage the whole project team to enable tasks and job descriptions for staff to be as diverse as practicably possible to instil variety and interest.

Other factors affecting motivation levels could be that education levels of construction operatives have historically tended to be relatively low and in some cases they have not been taught aspects of working independently, which in itself can affect their ability to motivate themselves (Tabassi and Bakar, 2009, p. 218).

Tabassi and Bakar (2009) also concluded that those construction companies which demonstrated a commitment to human resource development, motivated construction staff in the process, and reaped benefits associated with increased retention of staff and overall performance of employees. This is supported by Chen et al. (2003, as cited in Tabassi and Bakar, 2009, p. 214) who explained that: 'Having a well organised and effective human resource programme is one of the most important assets of a company, directly impacting its fruitfulness and long term viability as a company.'

Tabassi et al. (2011, p. 222) suggested that training can work in tandem with motivation and these two factors can work together to increase construction related productivity and prepare staff for organisational change and concluded that: 'Motivation can influence the willingness of an employee to follow the training programme to exert more energy towards the programme and to transfer what they have learnt onto the job.' Tabassi and Bakar (2009, p. 472) applied this view in their studies of construction companies in Iran who invested in training and career development for construction staff. Their findings concluded that such training and development fostered an integrated learning culture within construction organisations leading to higher levels of morale and motivation. Training for skilled construction staff 'on the job' involved job rotation to give individual transferable skills to different jobs and this increase in technical skills and training 'off the job' involving classroom lectures could be effective for developing technical and problem solving skills. Tabassi and Bakar also explained the concept of 'training motivation' whereby training can increase motivation levels. They concurred that this in itself can influence the energy and willingness that individuals will have for further training and development thus constituting a positive cycle. Cheng et al. (as cited in Tabassi and Bakar, 2009) advocated this approach and found 'training motivation influences trainees' training performance and transfer outcomes'.

Clearly there are many positive impacts that training, staff development and education can have on motivation levels of construction staff. Conversely, poor training and staff development can have potentially very harmful effects on motivation. This is evidenced in Kaming et al. (1998, p. 137) who concluded that poor management systems for coordinating construction staff and poor respect of them for one another can be a highly damaging factor on motivation levels. This would suggest that more focus through training on this element to encourage cooperation and planning between construction staff could be highly beneficial. In addition, Price et al. (2004, as cited in Dwivedula and Bredillet, 2010, p. 160) explained that organisational structures on construction projects are becoming increasingly flat and thus employee empowerment through training has become an important source of work motivation in recent times.

10.6 Cultural factors affecting levels of motivation for construction related staff

When considering construction industry employees, the importance of the culture of the construction site or contracting organisation can be sometimes underestimated as this can have a significant impact on job satisfaction of the workforce. A possible explanation for this could emanate from the insecure nature of contracting work given the one off nature of projects and the employment insecurity that can be derived from this. Within this context, perhaps construction staff are regarded by some as not requiring any form of motivational management and respond best to being simply given orders in what could only be described as a dictatorial manner. This analogy could explain why autocratic management styles with regard to communications and target setting for construction staff are so prevalent within the construction industry. It is not, however, to say that this form of management is not appropriate in all cases for managing construction staff. For instance, issues such as health and safety management, reliant on strict adherence to site rules and regulations, is of paramount importance and normally areas where there are clear warnings and potential disciplinary action for contravention. In such cases, site managers cannot simply rely on motivating construction staff to comply but must be adversarial and assertive to ensure regulations and the general safety of the construction workforce are maintained at all times.

What can sometimes be forgotten or underestimated is the need for construction staff to feel a sense of pride in their work and to feel a sense of teamwork in being part of a project. On these lines, McClelland (as cited in Brooks, 2006, pp. 76–89) presented a theory which promoted the need for staff to have a sense of belonging, esteem and achievement. This view is also supported by Elton Mayo (as cited in Deci, 1975, pp. 25–34). He explained that employees need to feel confident and comfortable within organisations in order to produce successful results and engage in teamwork. Accordingly he presented a theory which related motivation to human aspects. He advocated that wherever possible staff should enjoy their work and feel a sense of teamwork and personal satisfaction for undertaking their jobs. They should also feel valued within the organisation which in theory should incentivise them to work harder. This view is not, however, shared by Taylor (as cited in Deci, 1975, pp. 76–85) who conversely believes that motivation can be induced through more scientific means. In this regard, construction clients should encourage incentive related motivational measures. This could include giving staff adequate resources, introducing rewards for good performance and devising processes and systems for staff to feel more competent in carrying out their duties. Perhaps one argument against this approach could be that although it may give construction staff 'the tools' to do the job well it does not address the root cause of problems relating to poor motivation and morale issues. This in itself can create conflicts within construction organisations and more especially between subcontractor operatives and construction management teams. Brooks (2006, p. 45) addressed this issue and argued that conflict management policies and strategies within companies and institutions should be introduced into organisational management at all levels to avoid motivational difficulties.

Nesan and Holt (1999, pp. 122–129) explained that a powerful motivator of construction staff is the feeling of belonging to a team and this can be particularly effective where a team is given autonomy and authority to make their own decisions, raising levels of participation and communication. Tabassi and Bakar (2009, p. 474) reinforced this point and deduced that: 'Many employees are motivated when they are "empowered" and feel that their participation is important in making the company successful.' Tabassi and Bakar (2009, p. 474) also advocated that work environments should be created to maximise individuals' commitment and enthusiasm through motivational means. They explained that this can be achieved using incentives such as recognition and team belonging other than simply extrinsic motivators such as pay. Furthermore, Bellemare et al. (2010, p. 283) found that where workers experienced high levels of what they perceived to be alienation or pressure within organisational working environments, then this could have a negative effect on their productivity through lack of motivation.

Further studies on how organisational culture can affect motivation include Kaming et al. (1998, p. 134) who emphasised that 'de-motivating factors' are not necessarily the exact opposite of motivating factors and explained that most demotivating factors originate from apparent trivial causes, yet their impact is often significant. Their study, which looked at joiner, bricklaying and steel erection construction staff, found that disrespectful supervisors, lack of recognition and lack of cooperation amongst colleagues ranked as the main de-motivating factors. These in turn can lead to frustration, discontentment and thereby have a highly damaging effect on motivation levels leading to dissatisfaction and lower performance on site.

The culture of construction organisations can influence the responsibility assigned to construction staff on site. Lam and Gurland (2008, p. 1114) explained that where construction staff are trusted to do a job and given autonomy to carry it out, this leads to an increase in self-determined work motivation leading to improved job satisfaction. Conversely, they concluded the opposite effects where construction staff are given little if any autonomy or responsibility in a more controlled environment where they may be closely monitored and scrutinised. Construction clients predominantly acting as project sponsors should ensure that they pursue strategies designed to motivate their staff accordingly through responsibility and trust.

The culture of construction organisations and teams can also affect the degree of collaborative working between construction staff and management. Baiden et al. (2006, p. 21), however, concluded that integrated collaborative project cultures are seldom fully realised owing to barriers created by the short term duration of the majority of building projects. They explained that a lack of team working emanating from this could create a culture which alienates construction staff and makes them feel that they are working as individuals rather than team members. Accordingly, they recommended changes to the construction industry to encourage integration of teams working together following the principles of Latham (1994) and Egan (1998) to change cultures and improve motivation levels of construction staff. This recommendation was supported by Dwivedula and Bredillet (2010, p. 159) who advocated 'a complete cultural change within the construction industry' to give construction staff more autonomy and extensive training and thus incentivising creativity. They explained that this concept, when adopted by other industries such as manufacturing, and learning new skills through measures such as job rotation went towards motivating the workforce. This view was further reinforced by Reichers and Schneider (1990, pp. 88–98) who describe a 'shared perceptions approach' to organisational culture where a positive climate is developed when members of working groups are interacting with each other to attain common goals.

10.7 The use of financial incentives as a motivational management tool

The final factor for consideration in terms of its effects on motivation for construction clients is financial incentives.

Rose and Manley (2011, p. 765) attempted to explain where and why monetary rewards are used as follows: 'Financial incentives are typically used on construction projects to invigorate motivation towards above business as usual goals and provide the contractor with the opportunity for higher profit margins if exceptional performance is achieved.' Furthermore, they explained that financial incentives are common in the construction industry in terms of bonuses for trying to encourage motivation and commitment of construction staff, although they concede that there is little evidence to prove their effectiveness.

Olomolaiye and Price (as cited in Tabassi and Bakar, 2009, p. 474) apportioned much emphasis on money as a 'powerful motivator of construction employees'. MacKenzie and Harris (1984, p. 137) further supported this view and argued that money and earnings are the sole motivating factors for construction workers. This can be related to the psychological needs level of Maslow's needs theory. Further endorsement came from Kaming *et al.* (1998, p. 134) who found in their studies of construction workers in Indonesia that fairness of pay was the highest ranking motivational factor, which seems to again support the importance of financial gain. They did, however, also find that a culture that promotes good working relationships with workmates was ranked as the second most important motivator. In addition Olomolaiye and Price (1989, p. 283) conducted a survey of various construction groups involved in building projects in Nigeria. This rather contradicts that financial gain is the sole motivator for construction foremen and operatives with variables including good workmanship, job challenge and productive day gaining highest rank. Conversely, owners and project managers chose financial success as their prime job satisfier at the expense of the other variables, which rather suggests that these groups are more highly motivated by money than lower ranking groups including construction staff.

Herzberg *et al.* (1959, as cited in Tabassi and Bakar, 2009, p. 215) concluded that motivation of employees was linked to the absence or existence of certain job satisfiers. Examples of this include wages, conditions of work and holidays which could remove dissatisfaction but not greatly motivate people in their jobs. Story *et al.* concluded that financial incentives, if managed correctly and linked in with other motivational factors, can have beneficial results. Herzberg *et al.* (1959, as cited in Story *et al.*, 2009, p. 290) supported this view and concluded that intrinsic factors such as wanting to perform tasks well can lead to employees being satisfied at work whereas extrinsic factors, such as salaries and bonuses, typically lead to potential sources of dissatisfaction. When applied to construction staff, this could possibly explain why monetary rewards such as financial rewards for early completion may not greatly incentivise and motivate staff as would be expected in most cases. However, where they are seen to be removed from construction staff remuneration agreements, this could potentially have significant detrimental effects on demoralisation and motivation of the construction workforce. Schroder and Reich (as cited in Olomolaiye and Price, 1989) concluded that real construction motivation can only be derived from satisfying the 'higher needs' other than wages, working conditions and fringe benefits. In this way they argue that this can motivate them to achieve higher production which is a similar theory to McGregor's Y theory.

Tabassi and Bakar (2009, p. 474) referred to a 'participative approach' in addressing and developing good supervisor/subordinate relationships and motivation levels through the creation of cohesive work groups. They suggested examples of how this could be applied to construction staff, whereby they could be rewarded financially for suggesting alternative ways and means to improve the productivity, quality and timescales of their construction companies' building operations. Rose and Manley (2011, p. 765) concurred with this 'blended' approach and considered that financial incentives enhance motivational levels of construction staff to participate with construction related objectives, for instance early completion. They did, however, deem that there still need to be other factors instilled within the working environment such as unity, trust and fairness to make significant improvements to motivational levels of construction staff. Story *et al.* suggested that a balance can exist where construction staff can have both intrinsic and extrinsic motivations on a project. One example of this could be the desire to meet deadlines and the passion for doing a good job.

10.8 Summary and conclusions

To summarise the findings of this chapter, there are significant benefits from construction clients pursuing management initiatives on projects to motivate staff and therein lead to more successful outcomes. There has clearly in the past been an underestimation as to the benefits of motivational management and the potential implications for poor levels of construction staff motivation. The potential benefits could include:

- Improved construction staff retention
- More positive attitudes of construction staff
- Improved supervisor/subordinate relationships

- Increased productivity, and potentially higher quality standards
- Fewer conflicts and overall better communications on site

Given these clear benefits, this chapter has analysed and evaluated the factors that influence the motivational levels of construction staff. It has found that although financial incentives are undoubtedly important to construction staff they are by no means the sole mechanism for motivating the workforce. Perhaps the common misconception that construction staff are 'only in it for the money' has been generated from the unique environment that construction staff work within. As discussed, they are positioned at the bottom of the organisational hierarchy, their contracts are generally short term and they are working in an industry traditionally very competitive, especially in the current economic climates where financial incentives are seen as the 'quick fix' to meeting contract deadlines and targets.

Hopefully, this chapter of the book has contradicted the notion that construction staff are unworthy of requiring to be motivated by anything other than money and that attempting to incentivise and motivate them by anything else will be fruitless accordingly. In this regard, factors have been identified which construction clients should consider as they could have a significant impact on motivational levels of construction staff. These include improved communication strategies, consultation, training and development, organisational culture and opportunities. There is also a suggestion that the construction industry could benefit from adopting certain lessons learnt in other sectors where more emphasis has been placed on management strategies geared to raising levels of workforce motivation. Examples in this regard for new initiatives have been identified as job rotation and blue collar participation in strategic decision making.

Construction organisations have historically asked themselves questions related to whether they can afford to procure motivational management initiatives for construction staff. Furthermore, construction clients have traditionally distanced themselves from motivational initiatives under the misguided premise that it is not within their role to encourage project teams to motivate their staff. The real question, however, they should be asking themselves is can they afford not to? After all, how long can construction staff perform their duties and obtain good outcomes if not motivated and in some cases what level of performance can be expected where poor motivation levels are experienced?

This chapter of the book has emphasised why senior management, including construction clients, should pay special attention to motivational strategies. The justification for this is that low motivation levels within the workforce can otherwise result in low esteem, negativity, conflict, low productivity and frustration of construction staff. Conversely, if motivation is properly managed this can have very positive results as previously outlined. Clearly, motivation is not the only factor that affects productivity, and other factors, such as availability of resources within a particular situation, capability and ability to undertake a particular task, could be argued as being equally important.

Motivational management of construction staff could be used as a tool to enhance organisational performance and construction outcomes but is arguably as important to improve the mental wellbeing of construction staff in the process. Conversely, if motivational management is not practised, leading to construction staff becoming demoralised in some cases, construction operations may continue to work well with other 'negative' factors such as job security and fear of reprisal affecting performance on site. The overriding question is, however, how long can this be maintained in such scenarios? The overwhelming academic consensus would tend to suggest that a lack of motivation coupled with autocratic management styles can provide 'short term fixes'. However, it is generally recognised that poor morale and motivation of construction staff could, in the longer term, have extremely detrimental effects on the construction industry. Such implications could include poor employee retention resulting from construction staff seeking alternative employment elsewhere. This increase in mobilisation of staff could hinder internal development of construction project teams in terms of longstanding, knowledgeable staff having the right expertise and 'knowing the businesses' they work within. This in turn could increase costs for contractors and pass them on to construction clients when considering such factors as disruption, loss of continuity and training of new employees to fill vacancies left by demoralised employees.

This chapter has identified the general need for alignment of individual and organisational goals and the aim for win-win scenarios. In addition, it has highlighted the need on the part of construction clients for greater awareness of the importance of motivating construction staff. Accordingly, construction clients should champion a required change in cultures within the construction industry, adopting the recommendations of Latham (1994) and Egan (1998) to procure more integrated and motivated project teams. More emphasis by construction clients should therefore be placed on intrinsic factors rather than simply relying on financial incentives to raise motivation levels of construction staff. This by no means should involve simply removing financial rewards or any other extrinsic motivational measures that construction staff have traditionally relied upon, as this could negatively affect motivation. Instead construction clients could look wherever possible to blend intrinsic motivational measures into extrinsic measures. Examples of such initiatives could include financial rewards for good suggestions to improve the quality or buildability of a particular work element.

In the context of the complex and dynamic environment of the construction industry, and to realise benefits associated with motivational management there is clearly a change of culture and attitude required within the sector. To enable this change construction clients should consider the integration of the following recommended initiatives on their projects:

- Introduce 'bottom up' management initiatives to instil greater construction staff empowerment through responsibility and autonomy.
- Ensure that tasks for construction staff, through measures placed by construction clients on their contractors, are challenging, varied and interesting through initiatives such as job rotation. This will seek to reduce monotony, boredom, alienation and incentivise the workforce.

- Develop informal as well as formal lines of communication through regular feedback and appraisals.
- Give praise to construction staff as individuals and teams where appropriate and introduce recognition initiatives. Examples could include 'subcontractor team of the month'.
- Ensure that training and development of construction staff is undertaken and learning achievements recognised in individuals. This could be achieved through apprenticeships, other on the job training and through short term training courses off-site.
- Encourage construction staff to pursue career development through qualifications and industry accreditation.
- Develop and embed greater awareness of the importance of motivation for construction staff at senior management level.
- Break down communication and cultural barriers and treat construction staff as a more integral part of the overall project team and adopt more democratic rather than autocratic styles of management.

All these measures will hopefully contribute to provide a motivating work environment for construction staff to work within. In this way construction clients could reap great benefits not just for their respective organisations and the building projects that they procure but for the future of the construction industry at large.

References

Baiden, BK, Price, ADF and Dainty, ARJ (2006). The extent of team integration within construction projects. *International Journal of Project Management*, 24, 13–23.

Bellemare, C, Lepage, P and Shearer, B (2010). Peer pressure, incentives, and gender: an experimental analysis of motivation in the workplace. *Labour Economics*, 17, 276–283.

Brooks, I (2006). *Organisational Behaviour, Individuals, Groups and Organisations*. 3rd ed. Essex: Pearson Education Limited.

BusinessMate.Org (n.d.). *Victor Vroom's Expectancy Theory*. www.businessmate.org/Articles.php?ArtikelId=42. [Accessed 2nd February 2012].

Carnall, CA (1999). *Managing Change in Organisations*. 3rd ed. Hertfordshire: Prentice Hall Europe.

Challender, J (2017). *Collaborative trust in UK further education procurement strategies*. Unpublished PhD thesis. University of Bolton, UK.

Chen, LH, Liaw, SY and Lee, TZ (2003). Using an HRM pattern approach to examine the productivity of manufacturing firms. *International Journal of Manpower*, 24, 299–318.

Deci, E (1975). *Intrinsic Motivation*. New York: Plenum Press.

Dwivedula, R and Bredillet, NB (2010). Profiling work motivation of project workers. *International Journal of Project Management*, 28, 158–165.

Egan, J (1998). *Rethinking Construction: The Report of the Construction Task Force*. London: DETR.TSO.

Herzberg, F, Mausner, B and Synderman, B (1959). *The Motivation to Work*. New York: Wiley.

Kaming, PF, Olomolaiye, PO, Holt, GD and Harris, FC (1998). What motivates construction craftsmen in developing countries? A case study of Indonesia. *Building and Environment*, 33, 131–141.

Lam, CF and Gurland, ST (2008). Self-determined work motivation predicts job outcomes, but what predicts self-determined work motivation? *Journal of Research in Personality*, 42, 1109–1115.

Latham, M (1994). *Constructing the Team*. London: The Stationery Office.

MacKenzie, KI and Harris, FC (1984, 25–29 May). Money the only motivator. *Building Technology and Management*, 134–145.

Nesan, LJ and Holt, GD (1999). *Empowerment in Construction: The Way Forward for Performance Improvement*. Baldock, Hertfordshire: Research Studies Ltd.

Olomolaiye, PO and Price, ADF (1989). A review of construction operative motivation. *Building and Environment*, 24(3), 279–287.

Reichers, AE and Schneider, B (1990). *Climate and Culture: An Evolution of Constructs*. San Francisco: Jossey-Bass.

Rose, T and Manley, K (2011). Motivation toward financial incentive goals on construction projects. *Journal of Business Research*, 64, 765–773.

Steers, RM, Porter, LW and Bigley, GA (1996). *Motivation and Leadership at Work* 6th ed. Singapore: The Graw-Hill Companies Inc.

Story, PA, Hart, JW, Stasson, MF and Mahoney, JM (2009). Using a two-factor theory of achievement motivation to examine performance-based outcomes and self-regulatory processes. *Personality and Individual Differences*, 46, 391–395.

Tabassi, AA and Bakar, A (2009). Training, motivation, and performance: the case of human resource management in construction projects in Mashhad, Iran. *International Journal of Project Management*, 27, 471–480.

Tabassi, A, Ramli, M, Hassan, A and Bakar, A (2011). Effects of training and motivation practices on teamwork improvement and improvement and task efficiency: the case of construction firms. *International Journal of Project Management*, 30, 213–224.

11 Developing a construction client toolkit, as a practical guide for managing projects

11.1 Introduction

This chapter is designed to assist construction clients in the management of all stages of projects. Proformas and checklists have been specially developed for this chapter to improve the robustness of processes and for compiling important data on projects. These have been devised to ensure that critical aspects of projects, including statutory compliance and quality controls, are not compromised during the life of projects. In this way, it is intended as a practical guide to managing the processes and procedures linked to each aspect.

Following Driscoll's Three Elements of Reflection Model (Driscoll, 2000), the authors' reflective practice has led to reflective enquiries around construction clients and specifically leading projects. This triggered several questions which they sought to unravel. These included:

- Where is the evidence that justifies the idea that theory and practice need to come together in construction client awareness of leading and managing projects?
- Why is there a gap in provision around awareness of knowledge, training and guidance for construction clients?
- What improvements could be made to bridge the gap?

In consideration of the above questions, and from interviews with construction clients for this book, it has become apparent that some, especially those who lacked previous construction experience, felt ill-equipped with the basic skills required to successfully procure projects. Furthermore, these individuals articulated that they relied heavily on their design consultants to guide them through the necessary steps and processes. Their lack of expertise and training in professional practices around construction management led the authors to consider a 'toolkit' which could capture the various stages of projects and offer practical guidance at each stage. The intention of this guidance document would be predominantly to bridge gaps in knowledge and ensure that critical stages of projects are given sufficient consideration by clients.

11.2 Planning and devising the toolkit

As far as addressing the gap in knowledge, available guidance documents and general provision around raising awareness and competencies of construction clients, it was considered useful to develop an innovative and meaningful teaching and learning resource. The primary aim was to enhance the existing skills and awareness of clients as a knowledge base. It would seek to benefit clients when procuring projects and enable construction clients to improve their leadership and management skills to promote successful outcomes.

- Create a teaching and learning 'step by step' practical guide that will assist clients through the construction process and allow them to lead their project teams in a professional and proactive way.
- Familiarise construction clients with some of the necessary processes and governance requirements for projects to better prepare them for procuring projects.
- Close current gap between those existing education/training skills gained by construction clients and those required when procuring projects.

A project plan proposal form was prepared by the authors of this book which mapped out the aims and objectives of the practical guide, to assist in its creation. It was essentially designed to focus on the development of a simple and concise practical guide through the various procurement stages. This was intended as a guide and template to assist construction clients in undertaking building projects and leading them in a professional capacity. Its main purpose was intended to steer construction clients through a logical sequence and methodology for construction management processes from start to finish.

The basis of the practical guide or 'toolkit' in this chapter is to provide a reference tool to successfully procure projects and designed to give all the necessary information that construction clients require to enable them to comply with best practices. It could also provide construction clients with a basic familiarity and understanding on some of the procurement 'checks and balances' that should be considered. This will then hopefully benefit them in their projects and better prepare and equip them for some of the challenges faced.

11.3 Feedback and evaluation of the toolkit from the perspectives of construction clients

A consultation through interviews with construction clients was carried out to review the 'toolkit' and obtain their completed feedback on how useful they found the practical guide.

Positive feedback from all participants included the following:

- The toolkit is clear and unambiguous.
- The various checklists, templates and proformas contained within the toolkit allowed building information to be easily compiled throughout the life of projects.
- It promoted a logical and methodical step by step approach to carrying out building projects.
- As a practical guide it ensured that each stage of the construction procurement process was managed professionally and in a systematic manner to avoid 'checks and balances' and that governance considerations were not compromised.

Notwithstanding the above feedback, the advantage from an educational perspective is that the construction clients interviewed reported that they feel they are now more prepared and able to progress projects more confidently, using the information compiled within the toolkit. The guide was intended to steer them through the construction management processes, stage by stage, and allow them to comprehend and manage the various different aspects of projects. Furthermore, it is felt that the guide could be conducive as a teaching and learning tool for industry. In its existing simplistic form, it could potentially save a lot of time for clients in collating information more swiftly.

The following sections of the chapter articulate the various stages of projects and offer assistance to construction clients through templates, checklists and proformas as part of the aforementioned practical guide or 'toolkit'.

11.4 Ensuring and monitoring performance throughout the life of projects: general project directory and checklist

One of the important considerations for construction clients is to keep data on projects within one system or folder whereby information on projects can be recovered and updated easily. There should ideally be an overarching project checklist which should cover all stages of projects from inception to completion. This 'master checklist' is intended to capture other checklists at various stages of projects and therein be a useful management tool as projects progress. An example of a general project checklist is included in Figure 11.1.

11.5 The documentation that construction clients need to consider at pre-construction stages

There are many different documents that clients should source from their construction partners prior to works commencing on site. These are largely related to health and compliance and competency.

11.5.1 Contractor pre-qualification questionnaires for competency and compliance

Pre-questionnaires are sometimes commonplace. These are designed and implemented by construction clients to assess contractors' suitability prior to them

Project Title:					
Report No.		Report Date:		Time Elapsed:	

Overview and further items for discussion at the next progress meeting.

Programme — Status:

Accepted Programme Revision:		Date of the Last Issued Programme:	
Next Programme Revision:		Date to be Issued by:	

Pre-Construction		**Construction**		**Status Notes & Actions**
Start Date Pre-Con		Start Date Construction		
End date Pre-Con		End date Construction		
Planned Pre-Con completion		Planned Construction completion		
Variance:		Variance:	weeks	
PSC Period (calendar wks.)		Main Works Period (calendar wks) (Decant excluded)	weeks	

Budget — Status:

Pre-Construction (contract stage 1)		**Construction (contract stage 2)**		**Status Notes & Actions**
Original Pre-Con £ (end of RIBA 4)		Original Construction £ (RIBA 4-7)		
Forecast Pre-Con £ (end of RIBA 4)		Forecast Construction £ (RIBA 4-7)		
Variance:		Variance:		
Total MS (Contract Stages 1 + 2)		**Total Client (Contract Stages 1 + 2 + Client £)**		
Original total £ (RIBA 2-7)	£ -	Client affordability Gross £		
Forecast total £ (RIBA 2-7)	£ -	Forecast Gross £		
Nett Variance:	£ -	Previous agreed Gross £		
Cost plan ref		Gross variance to Client affordability:	£ -	Current Risk Allowance:

Design — Status:

				Status Notes & Actions
RIBA Stage		Client target gross area m2		
End of stage date		Forecast gross area m2		
Weeks left in this stage		Variance to gross area m2		
Feasibility Nett/Gross %		Proposed Nett/Gross ratio %		

Contract & Orders — Status:

				Status Notes & Actions
Pre-Construction Contact Status:		Contract Type:		
Main Contact Type:				
Draft Contract receipt date:		Anticipated Contract execution date:		

Early Warnings — Status:

				Status Notes & Actions
Number of EW Issued:		Number of EW open:		
Number of CE Issued:		Number of CE open:		

Description of Early Warning notices issued in the reporting period:

Figure 11.1 General project checklist

being included on tender lists. Construction clients normally adopt schemes such as the Construction Skills Certification Scheme (CSCS) and their questionnaires relate to pre-qualifying criteria for such affiliated accreditation. Contractors that are not part of these schemes must apply in writing and complete the relevant health and safety competency questionnaire. These are normally approved by the construction clients prior to commencing the tender processes.

On receipt of the contractor pre-qualification information, this should be forwarded to the relevant client officer who will add the details to the 'Contractor Compliance List' and advise the construction client if the tender action process can now progress. In some circumstances, it may be necessary for contractors to become members of a recognised scheme, due to the specialist nature of the works to be completed. These could include schemes such as the Considerate Contractors initiative/accreditation, linked to community liaison and ensuring neighbouring parties are not unduly affected by construction projects. An example of a pre-qualification competency questionnaire is contained in Appendix 1D.

11.5.2 Monitoring checklist required for document control and quality assurance processes

Following the tender processes leading to contractor selection, it is essential for construction clients to ensure that their contracting partners are checked to have the right compliance measures in place, prior to commencing work. Table 11.1 has been developed as a 'Monitoring Checklist' for construction clients or their project managers to ensure that all documentation is in place in this regard.

The checklist will serve construction clients to verify that their contractors have compliance measures in place and are fully inducted before being instructed to commence works. This is designed to ensure that the following governance provisions have been carried out or in place:

- All the necessary insurances (public and employer)
- Risk assessments
- Method statements
- Permits to work
- The Control of Substances to Health Regulations 2002 (COSH) assessments
- Completion of site and health and safety inductions
- Contacts and awareness of procedures for emergencies

11.5.3 Permits to work and making contractors aware of known hazards

Permits must be signed off and returned to appointed officers, nominated by construction clients, prior to work commencing and at completion. Typically, tasks and activities which may need to be authorised by a Permit to Work will be those that involve hazards and include:

- Asbestos removal
- Confined space entry
- Excavations
- Higher risk electrical work
- Hot work
- Work on fire alarm systems and emergency lighting
- Roof access/roof work
- Work at height.

Table 11.1 Monitoring checklist

Date :	Reviewer:	Position:
Time:	Contractor:	Lead name:
Permit type:	Location:	Area/room:
Brief description of job:		
Other permits:		

If any unsafe conditions are found, the work must be stopped and the Appointed Officer notified immediately.

PART 1 – CONTRACTOR COMPLIANCE	YES	NO	N/A
1 Has the Contractor had an induction in the last 12 months?			
2 Is the Contractor's safety scheme membership and liability insurance in date?			
3 Has the Contractor reviewed and signed the Asbestos Register?			
PART 2 – MONITORING of Work Authorization Certificate (WAC)/Permit to Work	YES	NO	N/A
4 Are job specific risk assessments and method statements available for review?			
5 Are the details of work to be undertaken clearly specified on the WAC/permit?			
6 Are the details of known hazards associated with the work listed on the WAC/permit?			
7 Are appropriate precautions listed on the WAC/permit?			
8 Is the operational time limit of the WAC/permit clear?			
9 Are other area or system activities cross-referenced correctly via the WAC/permit?			
10 Does the Contractor have the WAC on their person?			
11 Are copies of permits, certificates, COSHH assessments, etc. posted at correct locations?			
12 Are copies of WACs/permits, certificates, COSHH assessments, etc. legible?			
13 Are signatures traceable and legible?			
14 Are any other attachments, drawings, etc. held at the correct locations?			
15 Are users briefed on the WAC/permit, and have they acknowledged understanding of requirements?			

(*continued*)

Table 11.1 (Cont.)

16 Do people know what to do in the event of emergency?			
17 Does the Contractor have a copy of the employer's Contractor Compliance Handbook with them?			
18 Are isolations appropriate for the task, clearly specified on the permit, and correctly implemented?			
19 Are the right people aware of isolated equipment?			
20 Has the Area Supervisor/Building Manager been made aware of the work?			
21 Is there an Estates Notification of Work posted in the area?			
22 Is the work carried out in conformance with the WAC/permit/RAMS?			
23 Are the required control measures and personal protective equipment in place?			
24 Are tools and equipment suitable and in good condition?			
25 Are housekeeping standards satisfactory?			

Please use the following comments box to note down any issues/non-conformances and return the completed form to the Director of Estates.

COMMENTS

Item No.	
WAC/Permit Reviewer:	Name: Signature: Date:
Appointed Officer:	Name: Signature: Date:

11.6 Managing documentation and construction processes following the appointment of contractors

11.6.1 Project execution plan

A project execution plan (PEP) is the governing document that establishes the means to execute, monitor and control projects. The plan serves as the main communication vehicle to ensure that everyone is aware and knowledgeable of project objectives and how they will be accomplished. A PEP will typically contain information as set down in Appendix 1F.

11.6.2 Contractors' health and safety handbooks and codes of conduct

In addition to the above, it is good practice for construction clients to compile their own health and safety contractors' handbooks, which are normally tailored to their particular organisations and estates. These should incorporate important information on what contractors need to know about construction clients' sites and buildings and safety precautions that must be strictly adhered to. An example of a contractor's health and safety handbook is included in Appendix 1E. It is also common in these health and safety handbooks for contractors to have to sign up to an employer's code of conduct and site rules. These should ensure that contractors comply with the rules, regulations and procedures when entering into the clients' buildings and estates. Any shortcomings, in this regard, should be rectified immediately and the overall health and safety performance of a contractor reviewed regularly.

Construction clients should ideally appoint an officer within their respective organisations to carry out random site inspections to assess compliance with control measures employed by contractors. The frequency of the checks should be appropriate for the size, complexity, location and nature of projects. Clearly those projects that are potentially more complex and have the risk of causing more disruption to clients' businesses and staff should be managed more closely through such checks and inspections. This is particularly the case when contractors are working in 'live environments' and business continuity represents a major issue for clients. Clients have a duty of care to ensure that their buildings and estates are clean and safe for employees, visitors and members to enter into. Any accidents on their respective sites could lead to legal action being taken against client organisations for negligence and duty of care breaches.

11.7 The documentation and processes that construction clients need to consider in the construction phases of projects

11.7.1 The importance of monthly project reporting for construction clients

Construction clients frequently are responsible for more than one project which may be running concurrently at any time. These projects may be part of a programme of projects, and in some cases where there is considerable capital investment, a large transformational 'masterplan'. This could be particularly relevant to public sector bodies or large private sector clients embarking on regeneration schemes. This large scale development normally places undue challenges on construction clients in being able to have a 'full high level' perspective on a programme of projects. For this reason, it is important for construction clients to be provided with regular, preferably monthly, high level reports on progress of projects from their respective project teams. This should include an exceptive summary, progress against milestones, matters arising since the last report and any financial considerations. Furthermore, it should ideally incorporate a cost plan, reporting actual expenditure against planned, together with a 'dashboard' that could indicate project particulars including risks. An example of a cost plan is included in Table 11.2 and a progress dashboard template (blank) in Figure 11.2.

The monthly reporting should capture the current risk on projects, identified as part of the project execution plan (PEP) and report against them. Where there are risks, mitigation to control and reduce those identified risks should be documented. There should also be reporting on quality management in the build processes and a general overview of whether the building is performing to key performance quality standards and therein fit for purpose.

11.8 The documentation and processes that construction clients need to consider in the post-construction phases of projects

11.8.1 Managing handover: project handover checklists and test certification

On completion of projects, construction clients should have robust procedures to ensure that buildings for handover are safe to occupy. Arguably some of this responsibility clearly falls to their consultant project managers, but clients still have due diligence in terms of guaranteeing a safe and compliant working environment for their staff. In practice, many construction projects have been completed without the full array of 'checks and balances' in place to ensure legal and safe staff occupation. It would be extremely easy and commonplace for certain items related to such things as test certification and mechanical and electrical commissioning to be missed at the project handover stage. For this reason, the project handover/test certification checklist in Table 11.3 has been prepared to capture all the relevant items, normally required to be signed off prior to completion.

In addition to the test certification checklist, it is also good practice to complete a separate fire safety project completion checklist for health, safety and wellbeing. This covers such things as fire extinguisher installation, reinstatement of fire alarm detectors and call points, correct signage being in place showing new building layouts and ensuring clear/unimpeded fire escape routes. Table 11.4 has been prepared to illustrate a health, safety and wellbeing checklist. Construction clients' responsibilities should include ensuring that the most appropriate person, normally the project manager or health and safety manager, has signed off this checklist prior to taking handover.

Table 11.2 Example of a cost plan

2.1 CONSTRUCTION SUMMARY				
		This Report (11)	Previous Report (10)	Change + / (-)
1	**Contract Sum**			
2	Adjustment of PC and Provisional Sums (Appendix A)			
3	Adjustment of Day Works Allowance (Appendix B)			
4	Instructed Variations (Appendix C)			
5	Anticipated Variations (Appendix D)			
6	Contractual Claims (Appendix E)			
7	Project Risks (Appendix F)			
8	**Estimated Final Contract Value (Excl. VAT)**			
9	**Anticipated Over / (Under) Spend on Contract Sum [8-1]**			
10	**Client Directly Incurred Costs (Appendix G)**			
11	**Project Costs (Inc Fees & Excl VAT)**			
12	**Project Costs (Inc Fees & Inc VAT)**			
13	**Project Budget**			
14	**Variance Against Budget**			

2.2	COMMENTARY (Change in Period)			
(i)	This cost report includes the following movement in the reporting period:			
		This Report (11)	Last Report (10)	Change in Period
	Instructed Variations (Appendix C)			
	Anticipated Variations (Appendix D)			

Table 11.2 (Cont.)

2.2	COMMENTARY (Change in Period)			
Basis of Cost Report				
(ii)				
(iii)				
(iv)				
(v)				
Expenditure				
(vi)				
Current Issues				
(vii)				

11.8.2 Capturing lessons learnt on projects

Construction clients should ensure that there is a post-contract period of reflection to review what went well and not that well on projects. This would normally be conducted through workshops across the whole project team, including the end users. This reflection is designed to understand the lessons learnt which can be carried over to future projects, and therein improve future practices. In theory, without establishing those lessons learnt and what improvement measures should be adopted, the same mistakes or bad practices could continue from one project to another. Table 11.5 has been prepared to illustrate a proforma that could be used as an agenda throughout lessons learnt workshops. This lessons learnt proforma is split into three parts relating to the different stages of projects, namely pre-construction, construction phase and post-construction. Its primary purpose is intended to highlight areas for improvement and suggest improvement/mitigation measures to avoid the same issues arising in the future. To achieve full value from the lessons learnt exercise, a series of improvement measures should be formulated and an implementation strategy by which such measures can be integrated into projects. This normally involves reforming existing project management processes and procedures to capture the improvement measures at the relevant project stages.

STAGE	RIBA PLAN OF WORK 2013	RECEIVED	N/A
0	**Strategic Definition**		
1&2	**Feasibility and Concept Design**		
	Project Brief Form*	☒	☐
	Client Briefing Checklist		
	Total Forecast Expenditure		
3&4	**Developed/Technical Design**		
	Client Sign off Document*	☒	☐
	Statutory – Building Regulations Application	☐	☐
	Statutory – Planning Application	☒	☐
	Design Team Meeting Agenda		
	<u>Tender Stage Contract Documents</u>		
	Building Cabling Spec V6	☒	☐
	Car Parking Policy & Specifications	☐	☐
	UoS Contractor Handbook V1	☐	☐
	Location of Existing Wifi Points	☐	☐
	Location of Existing BMS Points	☐	☐
	Campus Map	☐	☐
5	**Construction**		
	<u>Pre-Construction</u>		
	Contractor H+S & Insurance Documents*	☒	☐
	Construction Phase Plan (RAMS) Received	☐	☐
	Fire Risk Information Pro-Forma (FRIP)*	☐	☐
	Notification of Building and Engineering Works*	☐	☐
	Asbestos - Notification of Building and Engineering Works*	☐	☒
	Laboratories – Notification of system of work received*		☒
	Pre-start Meeting agenda		
	Contact Directory Template		
	<u>Construction</u>		
	Progress Meeting Agenda Template		
	Progress Meeting Minutes Template		
	Security Weekend working Form*		
	Project Budget Report		
6	**Handover and Close Out**	RECIEVED	N/A
	Fire Safety Project Completion Checklist*	☐	☐
	Construction Waste Template*	☐	☐
	Final Schematics – Water Systems*	☐	☐
	Project Handover Checklist – Test Certification*	☐	☐
	O&M Manual Received*	☐	☐
	H&S File (Update)*	☐	☐
	As Built Drawings*	☐	☐
	Snagging Template		
	Additions to Security Systems/Provisions*		
7	**In Use**		
	Lessons Learnt*	☐	☐

*Please refer to stakeholder checklist for communication list

Figure 11.2 Example of a project dashboard template

Table 11.3 Project handover/test certification checklist

	DESCRIPTION	RECEIVED	N/A
1.00	**ELECTRICAL CERTIFICATES**		
	• Electrical Test Certificates	☐	☐
	• Fire Alarm Test Certificates	☐	☐
	• Emergency Lighting Test Certificate	☐	☐
	• Lift Commissioning Test Certificates & Manual	☐	☐
2.00	**MECHANICAL CERTIFICATES**		
	• Gas Suppression Installations	☐	☐
	• Natural Gas Installation Certificates / Schematic Drawings	☐	☐
	• HV Installations and Sub-stations Manual	☐	☐
	• Chlorination Certificate(s)	☐	☐
	• Air Cooling and Refrigeration Plant Commissioning*	☐	☐
	• Air Handling Units, Boilers and Plant Room Equipment	☐	☐
	• Fume Cupboards and Local Extract Ventilation	☐	☐
	• Building Management System		
3.00	**ITS CERTIFICATES**		
	• Data Installation Test Certificates	☐	☐
	• AV Installation Commissioning	☐	☐
4.00	**OTHER**		
	• Lightening Protection Commissioning Certificates	☐	☐
	• Fire Door and Fire Stopping Certificates	☐	☐
	• Access Equipment Test Certificates	☐	☐
	• Energy Certificates – EPC and DEC	☐	☐

Table 11.4 Fire safety project completion checklist

Project Officer	
Project Title	
Project Code	
Completion Date	

Have all fire alarm detectors identified on the FRIP been reinstated?			Y/N
Have fire extinguishers been replaced/installed as per the Fire Strategy document?			Y/N
Has the following fire signage been replaced using photo-luminescent signage:	Manual call point		Y/N
	Fire extinguishers		Y/N
	Do not use lift		Y/N
	Door override buttons (green break glasses)		Y/N
Have Fire Action notices been replaced?			Y/N
Have fire escape routes been maintained and signage provided (these must be illuminated in licensed premises)?			Y/N
Have refuge points been resigned and communications maintained?			Y/N
Have fire doors been maintained to be opened with one action and both leafs available (bolts must not be fitted to these)?			Y/N
Has fire stopping been installed and checked on any penetrations to walls?			Y/N
Have building plans been updated showing new layouts, etc.?			Y/N

Table 11.5 Lessons learnt proforma

Stage 1 – Prior to commencement of the works	YES	NO	NOT APPLICABLE OR UNSURE
1 Did we make effective initial contact with you?			
Comments:			
2 Did we communicate with you effectively?			
Comments:			
3 Were we supportive of the aims of your department?			
Comments:			
4 Did we explain the project budget in appropriate detail?			
Comments:			
5 Did we fully understand your department's functions & critical requirements?			
Comments:			

Stage 2 – During construction/refurbishment	YES	NO	NOT APPLICABLE OR UNSURE
6 Was the planning and phasing of work effective?			
Comments:			

7 Were you involved in decision making where appropriate?			
Comments:			
8 Did we respond to your requests/queries in a timely manner?			
Comments:			
9 Did we give you appropriate early alerts to potential problems?			
Comments:			
10 Was any disruption of your functions/services anticipated?			
Comments:			

Stage 3 – Completed construction/refurbishment	YES	NO	NOT APPLICABLE OR UNSURE
11 Did the completed work comply with the agreed brief?			
Comments:			
12 Did the completed work meet your accommodation needs?			
Comments:			

13 Did the completed work provide an environment that satisfies its users?			
Comments:			
14 Did we deliver a completed project that met your expectations taking into account budgetary restraints?			
Comments:			

11.9 Conclusion

In general terms the information included in this chapter could become a useful, simple and innovative practical guide for construction clients, especially those with little or no previous experience in managing projects. It could offer them a useful management tool, as relative beginners to the task of carrying out complex building projects, and hopefully narrows the gap between their existing knowledge base and those skills required for successful construction management. The various proforma and templates, tailored to what is believed construction clients require, should form the impetus for improved practices as checklists to ensure critical information and processes are strictly adhered to.

The opportunities for the guidance contained in this chapter could support future development for professional practice and in an education context provide a useful teaching and learning resource. This guidance as part of the toolkit presents a more effective and efficient means of collating building information and a step by step route map through complex procurement stages and processes. It has the potential to become integrated into Building Information Modelling (BIM) systems which client organisations may wish to implement. This technology could be networked and downloaded onto a software system that could effectively prepare a report automatically. This would make the whole process so much more efficient and greatly reduce the resources and time currently expended by construction clients' staff in preparing various reports. This is particularly relevant for client companies seeking alternative ways to reduce costs and become more competitive accordingly.

Reference

Driscoll, MP (2000). *Learning for Instruction*. 2nd ed. Boston, MA: Allyn & Bacon.

12 Reflections, overview and summary of key points of Part 1

12.1 Overall summary and recommendations

The overarching aim of the chapters in Part 1 of this book is to provide a factual client 'how to do it' guide or 'toolkit' for procuring more successful project outcomes. It was intended that this practical guide for clients can develop into a common due diligence framework on how to initiate, procure and manage construction projects and developments. In this regard, the book has investigated the current arrangements that exist within the global construction industry, to create a more comprehensive understanding of the problems of client knowledge, interface and client involvement/integration within project teams.

Clients are important to the construction processes as they are normally the creators and funders of projects and as such the drivers for their developments. The ultimate goals of projects should be geared around clients' requirements in terms of their aspirations, ambitions, visions, aims and objectives. Accordingly, it is of paramount importance that the requirements of clients are fully articulated and understood by themselves and others involved in their projects. The 'voice of the client' (or client's requirements) includes the collective wishes, perspectives and expectations of the various components of the client body. These requirements describe the facility that will satisfy the client's objectives (or business needs). Client requirements constitute the primary source of information for a construction project and, therefore, are of vital importance to the successful planning and implementation of a project.

The degree to which client organisations are 'complex' will vary according to the composition, size and nature of their respective organisations. They vary from being small family run businesses to large global organisations or public sector bodies. What can be particularly challenging for projects is where very complex organisations have construction clients who are regarded as being relatively inexperienced. This represents a potentially difficult situation where the degree of risk on the project could be substantial as a consequence of the 'client voice' not being properly established. Furthermore, construction client organisations are diverse and come in many 'shapes and sizes'. Examples of these organisations have been given in the book as funding bodies, neighbourhood organisations, local authorities and community groups. In addition, they may comprise small family businesses or charities. One of the common mistakes that construction clients make is to not include those informed 'end users' in the early stages of consultation. Traditionally this has led to some occupiers of new buildings not feeling 'bought in' to the delivery of projects, and any contributions they may have had not materialising.

The construction industry has emerged as a bespoke project based industry where there are many different characteristics to other industries such as manufacturing. This is largely due to construction projects being nearly always unique and 'tailor made' to suit clients' individual requirement. Arguments have emerged that mistrust has been inherent within the UK construction industry for a long time between all parties including clients and contractors they employ. This may have stemmed from the traditional forms of procuring construction work which over recent years has been blamed for achieving low client satisfaction levels, poor cost predictability and time certainty. Arguments have been raised in Part 1 of this book that 'collaborative working' or 'partnering' offers a more suitable alternative for construction clients in procuring more successful outcomes for projects. Projects which had applied principles of both Latham and Egan in the use of collaborative procurement methods, have led to significant improvements in client satisfaction, cost predictability, safety and time predictability. In this context, such a partnering based approach has been proven to encourage cooperation and trust.

The book has referenced current academic literature on theories relating to leadership and applied these to the roles of construction clients. As mostly leaders on projects, it is essential for these individuals to understand what makes a good leader, in order to perform their roles and motivate others for achieving successful outcomes. The book has highlighted the 'leadership identity development and progression process' as applied to construction clients. In the context of securing successful outcomes, it has described how the model is dependent and influenced by individual differences, cognitive capacity, personality and temperament, personal identity, personal values and emotional intelligence, driven by cultural context and personal experience.

The importance of governance and adherence to financially robust procurement and approval processes for construction clients has been articulated in this part of the book. It is imperative that construction clients understand their governance and legal roles, responsibilities and authorisation levels when leading their project teams. Furthermore, it is important for construction clients to have robust decision making processes in place within their respective organisations whilst ensuring governance procedures are maintained. In considering the approval processes for projects through the individual boards, it is commonplace for robust and financially rigorous business cases to be prepared to support the business ventures.

The importance for construction clients of the processes around selection and appointment of their contracting partners has been discussed widely in this part of the book. It has articulated that having the right main contracting partners on board is arguably one of the most important aspects for construction clients. In this regard, it is imperative for clients to deploy robust selection processes for selecting the most appropriate construction contractors to realise benefits of partnering through encouraging pro-activity, building teamwork ethos, employing lateral thinking and exploring/devising alternative building solutions. Construction clients should develop and implement project specific and objective pre-qualification processes and ensure that any information submitted by contractors is validated by contacting cited referees, and if possible other past clients not identified by contractors.

The book has identified barriers and obstacles to trust generation within construction strategies. These mostly revolve around commercial issues and traditional adversarial behaviours of project teams. Improvement measures to address such issues include more focus on collaborative working as part of construction management curriculum, awareness of benefits of partnering philosophies through CPD and use of more collaborative management tools. Perhaps the biggest challenges for construction clients remain around culture change within the construction industry and achieving longer term collaborative relationships between their partnering organisations. Other challenges have been identified relating to the general lack of any incentivisation for contractors to be more collaborative; clients have still not fully embraced the concept of collaborative working themselves. In addition, the book has highlighted an apparent lack of trust in project procurement strategies. Accordingly, construction clients should be aware of this major obstacle and consider strategies linked to partnering and building trust building strategies. These have been established by this book as having a strong influence on raising trust levels on projects, especially when linked with partnering procurement. Examples of successful initiatives for construction clients could include strategies around incentive provisions, workshops, CPD, collaboration management systems, senior management commitment, open and joint evaluation policies and improved communications.

Construction clients have been identified in the book as requiring different qualities especially focused around social skills, management and interpersonal skills. Management of projects from the perspectives of construction clients has become sophisticated in recent years with development of skills linked to achieving projects' objectives. There appears to be one particular skill that is common and applied to all and this relates to human skills. The challenge is to understand that the two different types of skills discussed at length in this book, namely technical and human, are both important and required for successful outcomes. More importantly, construction clients need to strike the right balance between the skills where and when needed, especially given that no project or project environment is similar or the same. Accordingly, success may not be achieved if the right balance of these skills is not present.

Construction clients, as leaders in the procurement of projects, should be driving the process of cultural change to improve the reputation of the industry. In this pursuit, the book has called for greater awareness from clients' perspectives of the importance of ethics, codes of conduct and governance and regulations in avoiding bad practices. Construction clients should be leading the way for a cultural change in the industry to train, educate and motivate construction individuals and organisations in what professional ethics entail, measures to ensure compliance and the benefits that they can bring for the sector.

There are significant benefits from construction clients pursuing management initiatives on projects to motivate staff and therein encourage more successful outcomes. There has clearly in the past been an underestimation as to the benefits of motivational management and the potential implications for poor levels of motivation in the construction industry. The potential benefits could include improved construction staff retention, potentially higher quality standards and increased productivity. Construction organisations have historically asked themselves questions related to whether they can afford to procure motivational management initiatives for construction staff. The book has, however, articulated that they should be asking themselves if they can afford not to? Furthermore, construction clients have traditionally distanced themselves from motivational initiatives under the misguided premise that it is not within their role or responsibility. This part of the book has hopefully emphasised why senior management, including construction clients, should pay special attention to motivational strategies. The justification for this is that low motivation levels within the workforce can otherwise result in low esteem, negativity, conflict, low productivity and poor construction outputs. Conversely, if motivation is properly managed this can have very positive results as previously outlined. Clearly, motivation is not the only factor that affects productivity and other factors, such as availability of resources within a particular situation, capability and ability to undertake a particular task, could be argued as being equally important.

The book has designed a 'toolkit' to assist construction clients in the management of all stages of projects. Proformas and checklists have been specially developed in Chapter 11 to improve the robustness of processes and for compiling important data on projects. These have been devised to ensure that critical aspects of projects, including statutory compliance and quality controls, are not compromised during the life cycle of projects. In this way, it has developed as a practical guide to managing the processes and procedures linked to each aspect. The information could become a useful, simple and innovative practical guide for construction clients, especially those with little or no previous experience in managing projects. It could offer them a useful management tool, as relative beginners to the task of carrying out complex building projects, and hopefully narrows the gap between their existing knowledge base and those skills required for successful construction management. The various proforma and templates, tailored to what is believed construction clients require, should form the impetus for improved practices as checklists to ensure critical information and processes are strictly adhered to. The opportunities for the guidance contained in this toolkit could support future development for professional practice and in an education context, and present itself as a valuable and useful teaching and learning resource. Furthermore, the toolkit could represent a more effective and efficient means of collating building information and a step by step route map through complex procurement stages and processes.

Appendix 1A: Project proposal (Gateway 1)

The writings in italics are for guidance only. Please delete them when you have completed your Mandate.

*This form is an initial summary of the reason(s) for the proposed project. It is required that sufficient information should be provided to enable decisions to be made on whether the proposal can proceed to the next stage of the Business Case approval process. The form is to be completed principally by the Project Manager in consultation with the Project Sponsor. A guidance document is also available and may be accessed **from this link**.*

STRATEGIC-LEVEL DEVELOPMENT (i.e. initiated at institutional level)		YES	NO
		☐	☐
School/Service:			
Project Title:			
Programme:	*If the project is a part of a Programme*		
Funded by:	*Please indicate the main funding source for the project*		
Project Sponsor:			

(1) STRATEGIC FIT / RATIONALE

Please explain the project in context by providing the reasons for wanting the project. Include also a brief statement of how the project fits into current Client's strategy.

(2) SCOPE & OBJECTIVES

List what the project aims to achieve in terms of objectives, and outline the scope of the project as project deliverables. Include anything which is excluded from the project (out-of-scope).

(3) SUMMARY OF KEY BENEFITS

List the benefits that will be realised by achieving the scope of work. Ensure that you emphasise the key benefits, for instance by highlighting the income streams that will be grown, cost/efficiency savings, services that will be refined, etc. Include also how the benefits will be measured/demonstrated.

(4) PROPOSED TIMESCALES

Provide a high level timescale for project execution and implementation. In particular include the estimated time for completing the project proposal, business case and key milestones. This is indicative only but should be based on what is currently believed to be achievable.

(5) STRUCTURE OF GOVERNANCE

List the key stakeholders for the proposed project. Also indicate what consultation is needed or has already taken place.

(6) PROJECT DEPENDENCIES

List any other project whose success depends on the successful delivery of the proposed project, or other externalities on which the proposed project depends on.

(7) PROJECT RISKS

Outline the major risks which will need to be managed for the project.

(8) PROJECT COSTS

Provide a high level summary of the likely costs for the project to provide an idea of scale and affordability. Include also the cost type, e.g. fixed and variable. This is indicative only but should be based on what is currently believed to be achievable.

(9) RESOURCES REQUIRED

List the key resources that are required to complete the project proposal stage of the business case approval. These include both financial and human resources, and indicate whether internal or external capability will be utilised.

Appendix 1B: Business case (Gateway 3) template

BUSINESS CASE (GATEWAY 3) TEMPLATE

PURPOSE OF DOCUMENT

The purpose of this document is to set out in detail the justification for the undertaking of a project based on the strategic objective, cost of the development and the identified benefits the Client will see post implementation. The business case should clearly say why the effort and time will be worth the investment. It should also provide reassurance that the project will be managed throughout the implementation stage and how the benefits identified will be measured. Provide reference to the approved Project Mandate and Project Proposal documents.

PROJECT DETAILS

Project Name	
Project Sponsor	
Project Manager	
Proposed Start Date	
Proposed Completion Date	

DOCUMENT CONTROL

Version No.	Date	Details of Change	Authors

Table of Contents

PURPOSE OF DOCUMENT
EXECUTIVE SUMMARY
SECTION 1: STRATEGIC CONTEXT: THE CASE FOR CHANGE
    1.1 Organisational Overview
    1.2 Project Background / Rationale
    1.3 The Drivers for Project
    1.4 Risk and Impacts of not continuing with the Project
SECTION 2: STRATEGIC CONTEXT: THE SOLUTION
    2.1 Project Purposes
        2.1.1 Project Goal
        2.1.2 Project Objectives
        2.1.3 Project Outcomes
    2.2 Project Scope
    2.3 Critical Success Factors (CSFs)
    2.4 Strategic Fit
    2.5 Project Benefits and Benefits Realisation Plan
        2.5.1 Main Benefits Criteria (if applicable)
        2.5.2 Main Project Dis-benefits
        2.5.3 Benefit Realisation Register
    2.6 Costs
    2.7 Timeline
    2.8 Project Constraint
    2.9 Project Dependency
    2.10 Project Assumptions
    2.11 Project Risks
        2.11.1 Risk Assessment
        2.11.2 Risk Register
SECTION 3: SUSTAINABILITY IMPACT ANALYSIS
    3.1 Sustainability Impact
    3.2 Sustainability Action Plan
SECTION 4: OPTIONS APPRAISAL SUMMARY
    4.1 Summary of Options Considered
    4.2 Option Evaluation Criteria
        4.2.1 Cost-Benefit Analysis
    4.3 Rationale for Preferred Option
SECTION 5: IMPLEMENTATION PLAN
    5.1 Project Governance
    5.2 Project Management Strategy
    5.3 Project Milestones Plan
    5.4 Risk Management Plan
    5.5 Change Control Strategy
    5.6 Project Assurance Strategy
APPENDIX
PROJECT REVIEWER DECISION TO PROCEED

The writings in italics are for guidance only and should be deleted once you have completed your form.

EXECUTIVE SUMMARY

The executive summary is a high level summary of the Business Case. It should include pertinent information to convey to the reader an understanding of the whole document in a clear and concise manner. It may be organised using the Business Case Section headings

SECTION 1: STRATEGIC CONTEXT: THE CASE FOR CHANGE

This section describes the current state of affairs and explains the need for the project from the perspective of the sponsoring Department.

1.1 ORGANISATIONAL OVERVIEW

Please provide an overview of the sponsoring Department. The overview should include Strategic Goals and Objectives, Current activities and Services, a high level organisational structure and key stakeholders and clients. Existing capacity (financial and human resources may be included at discretion.

1.2 PROJECT BACKGROUND / RATIONALE

Please describe the background to the project as well explain the current state of affairs including the issues or opportunities which the project seeks to address (Business need). Include any changes since the submission of the Project Proposal (Gateway 2).

1.3 THE DRIVERS FOR PROJECT

Please identify both internal and external drivers for the project and link them to the business need.

1.4 RISK AND IMPACTS OF NOT CONTINUING WITH THE PROJECT

List the consequences of not proceeding with the project

SECTION 2: STRATEGIC CONTEXT: THE SOLUTION

This section explains the nature and purpose of the project and how the project can solve the problems identified in Section 1 of the Business Case. It describes the project deliverables, benefits, alignment with the overall strategic objectives of the Client, costs, risks and other considerations which can affect the successful delivery of the project.

2.1 PROJECT PURPOSES

The project purposes subsection describes what the project is all about and the expected outcome from the project.

2.1.1 Project Goal

Please provide a simple statement on what the project is about.

2.1.2 Project Objectives

Please break down the project Aim into specific objectives to enable the achievement of the stated Aim. Objectives should be SMART (specific, measurable, achievable, realistic, time-bound).

2.1.3 Project Outcomes

Please describe at a high level the expected project outcome(s). This is not the project output but rather the desired state of affairs, in other words, what the project is intended to achieve. The project outcomes provide a link between the project objectives and the benefit to be derived from the project.

2.2 PROJECT SCOPE

Please describe the boundary (in-scope, and out-of-scope) of the project. The description should be in terms of identified business needs and any changes since submission of the Project Proposal. This may be considered along a continuum that may include the following:

Project Scope Boundaries

Project Boundary	Included	Excluded
Minimum scope: Essential requirement/project outcomes		
Intermediate scope: Essential and desirable project outcomes		
Maximum scope: Essential, desirable and optional project outcomes		

Consider if there are overlaps with other projects. Further information is provided in the Business Case Guidance document.

2.3 CRITICAL SUCCESS FACTORS (CSFS)

Please list and quantify the CSFs for the project against which the successful delivery of the project will be assessed as well as used in the evaluation of options. Further information is available in the Business Case Guidance document.

2.4 STRATEGIC FIT

Please demonstrate how the project fits within the broader Client strategic context and its contribution towards the achievement of Client priorities by completing the Table below.

Rating legend			
Blank	**Low**	**Medium**	**High**
Not applicable to the priority	Possible or uncertain contribution to the priority	Probable contribution to the priority	Clear and valuable contribution to the priority

Benefit cross reference
In this box identify which benefits A, B, C, D, E, F, G, H, I or J (Section 2.5) will impact upon which strategic objective.

Key measures
Measures to be adopted to measure that the project delivers the identified benefits. These can be existing KPIs, other key measures, customer satisfaction indices or a newly introduced measure specific to the project.

Project Strategic Fit

Sub Strategic Priorities	Key benefits and measures	Rating		
		H	M	L
Growth and Diversification:	Key benefits •			
	Key measures •			
Experience: Development with industrial partnership	Key benefits •			
	Key measures •			
Experience: Provide opportunity to develop as active partners, become self-aware and responsible	Key benefits •			
	Key measures •			
Experience: Opportunities to develop skills, attitudes, personal competencies and attributes	Key benefits •			
	Key measures •			

Sub Strategic Priorities	Key benefits and measures	Rating		
		H	M	L
Research and Enterprise: High quality research with both national and international impact; increased research and commercial income	Key benefits •			
	Key measures •			
Research and Enterprise: Development of new and existing institutional and industrial partnerships	Key benefits •			
	Key measures •			
International Priorities: Increase in growth opportunities	Key benefits •			
	Key measures •			
International Priorities: Reputational risk management and enhancement of the Client's international profile	Key benefits •			
	Key measures •			

2.5 PROJECT BENEFITS AND BENEFITS REALISATION PLAN

2.5.1 Main Benefits Criteria (if applicable)

Please describe in detail the main benefits associated with satisfying the scope of the project. Where possible this could be expressed in monetary form but bear in mind that some benefits may be qualitative in nature.

Project Benefits

PROJECT BENEFITS						
Type	Tick as appropriate		✓	Please explain and include value as applicable	One time value (£)	Annual Value (£)
Financial Benefits	A	Cash releasing, e.g. cost avoidance				
	B	Non- Cash Releasing, e.g. Staff time saved				
	C	New Income				
	D	Additional Income				

Appendix 1B 161

PROJECT BENEFITS					
Non-Financial Benefit	E	Strategic Fit	Cross reference with Section 2.4 to identify the strategic fit.		
	F	Competitive Advantage			
	G	Competitive Response			
	H	Operational improvement/ management			
	I	Others, e.g. Staff moral			
Risk Avoidance	J			One time value (£)	Annual Value (£)
Additional Information					

2.5.2 Main Project Dis-benefits

Please describe any dis-benefit associated with the project.

Project Dis-Benefits

Description	Stakeholder Group (as applicable)	How and When
Dis-benefit 1		
Dis-benefit 2		
Dis-benefit 3		

2.5.3 Benefit Realisation Register

*Please complete the project benefits register as an appendix to this document. The register sets out the responsibility for the delivery of benefits, their measurements and time frame for realisation. The benefits register template may be downloaded **from here**.*

2.6 COSTS

The costs of the project should be estimated using the standard cost estimation practices. Please indicate the person responsible for estimating the project costs and the standard method used. While a summary of the estimated costs of the project is required to be provided in the main body of the Business case, a detailed breakdown of costs is to be also provided as an appendix to the document.

Summary of the Estimated Project Costs

	Year 1	Year 2	Year 3	Year 4	Year 5
Estimated Capital Costs (as applicable)					
Hardware					
Software					
Others					
Total					
Estimated Revenue Costs (as applicable)					
Staff Costs – External					
Staff Costs – Internal (backfill)					
Maintenance charges					
Other Non-Pay Costs, e.g. contingencies, inflation costs					
Total					
Please provide advice on the following: 1. Anticipated source of funding for the project 2. Impact on Client's borrowing / gearing 3. Any applicable funding deadlines 4. Possible cash savings 5. Possible additional income 6. The impact on staff 7. Procurement issues / decisions					

2.7 TIMELINE

Please provide an overview of the timescale for the project. Highlight deadline dates associated with key milestones.

2.8 PROJECT CONSTRAINT

Please provide a summary of the constraints within which the project is expected to be undertaken. Also highlight any changes since the submission of the Project Proposal.

Summary of Project Constraints

Project	Brief description of Constraint
Constraint 1	
Constraint 2	
Constraint 3	

For each of the above, any changes since the submission of the Project Proposal should be highlighted.

2.9 PROJECT DEPENDENCY

Please list all project dependencies (if applicable) which need to be managed for the entire duration of the project. Any changes since the submission of the Project Proposal should be highlighted.

Summary of Project Dependencies

Project	Type of dependency	Brief summary of linkages
Project A	Dependent on this project	Requires delivery of output x, y and z from this project
Project B	This project depends on B	Requires project B to deliver products a, b and c

2.10 PROJECT ASSUMPTIONS

Please list and describe all assumptions, for instance with scope definition, benefit realisation, cost, etc., and the potential impact they could have on the project if not addressed. The Table below could be used for the purpose.

Summary of Project Assumptions

Number	It is assumed that:	Effects on Project	Reliability Level: High/Medium/Low
Assumption 1			
Assumption 2			
Assumption 3			

N.B.: Assumptions, constraints and dependencies are significant sources of risks and issues for any project. It is therefore important that the relationship among them are consistently made clear. The information is both useful in the planning stage of the project and during options analysis. They also have a big impact on benefit realisation if proved to be false or unreliable.

2.11 PROJECT RISKS

Risk Identification
Please identify the potential risks to the project and assess them in terms of their probability of occurrence and impact. Consider risk for the entire lifecycle of the project including both project delivery risks and project outcomes risks. More information is available in the Business Case guidance document.

Risk Identification for the Project

Project Delivery Risk				
Risks ID	Risks	What is at risk	Source (How can the risk occur)	Impact (what would the effect of the risk)

2.11.1 Risk Assessment

Please assess the identified risk in terms of their impact and probability of occurrence. A risk assessment matrix (RAM) should also be produced to prioritise the identified risks. Based on your assessment, please categorise the assessed risks as minor, medium and/or major.

You may consider producing the table below in a landscape layout.

Risk Assessment for the Project

Risks ID	Probability (P)	Impact (I)	Risk Score (PXI)	Risk Category	Tolerance Rating*

* Each risk assessed as major or medium should be rated as one of the following: Acceptable, Unacceptable or Unknown. Please refer to the **Business Case Guidance document for** additional information.

2.11.2 Risk Register

*Please complete the initial Risk Register as an appendix to this document. The Risk should describe the attributes of each major and medium risks as well as assign responsibility for the management of identified risks. The information in the risk register for each risk should include, a risk ID, risk statement, impact/probability rating, risk prioritisation level. Mitigation action, risk status and risk owner. A template is provided and can be downloaded from **this link.***

SECTION 3: SUSTAINABILITY IMPACT ANALYSIS

This section describes the impact the project will have on project stakeholders (including the Client) and on sustainability.

3.1 SUSTAINABILITY IMPACT

Sustainability remains the key enabler for the Client Strategic plans. Please use this section to summarise the impact of the project on sustainability under the following subheadings:

- **Financial Sustainability**
- **Environmental Sustainability**
- **Social Sustainability**

3.2 SUSTAINABILITY ACTION PLAN

If applicable please state what actions are being put in place to remove/mitigate any potentially negative impact on sustainability

SECTION 4: OPTIONS APPRAISAL SUMMARY

This section assesses the various options that could address the identified business need including the preferred option. The business case is written on the basis that the preferred option is to be adopted.

4.1 SUMMARY OF OPTIONS CONSIDERED

The preferred option for the project is contained within the body of the Business Case. A summary of other options considered are provided below.

Please list and summarise all the options considered to meet the identified business need including the do nothing option. The do nothing option should describe what will happen if the project did not go ahead.

For each option identified, please carry out a SWOT analysis of potential options to narrow down your option to three possible options.

Options	Strengths	Weaknesses	Opportunities	Threats
Option 1				
Option 2				
Option 3				
Option 4				
Option 5				

Summary Option Ranking	
Option 1	Possible
Option 2	Possible
Option 3	Possible
Option 4	**Discounted**
Option 5	**Discounted**

4.2 OPTION EVALUATION CRITERIA

The Table below is designed to assist in the assessment of the possible options by considering the financial implications and the non-financial contributions of the three possible options identified above. Please complete it and rank your options based on performances on the indicators.

Financial Appraisal	Option 1	Option 2	Option 3
Financial Appraisal			
Total Project Cost			
Total Additional Ongoing Costs			
Quantifiable Cash Benefits			
Net Present Value (NPV)			
Non-Financial Appraisal (Critical Success Factors)			
Project Outcome / Benefit			
Contribution to Client KPI			
Improve Space Utilisation			
Option Ranking	3rd	2nd	1st

4.2.1 Cost-Benefit Analysis

A further cost-benefit analysis may be conducted at discretion to further assist in the selection of a preferred option. More information is provided in the Business Case guidance document.

4.3 RATIONALE FOR PREFERRED OPTION

Please explain the reasons for choosing the preferred option as well as for discounting other options. Consider this in the light of the result of the evaluation criteria.

SECTION 5: IMPLEMENTATION PLAN

This section describes the project management methodology and approach to be adopted for the delivery of the project.

5.1 PROJECT GOVERNANCE

Please summarise the key roles and responsibilities for those involved in the delivering of the project including the reporting arrangement.

Appendix 1B 167

Also advise whether a steering committee or Project Board will be required. If applicable please outline what representatives will be required.

Summary of Project Governance

Roles	Name of Personnel	Responsibility (What each have to deliver to the project)
Steering Committee / Project Board		
Project Sponsor		
Project Manager		
Project Team		
Project Management Office		

5.2 PROJECT MANAGEMENT STRATEGY

Please summarise the project management approach to be adopted to managing the project.

5.3 PROJECT MILESTONES PLAN

Please describe the key things that need to be delivered in order to meet identified deadlines. More information is available in the Business Case guidance document.

5.4 RISK MANAGEMENT PLAN

Please summarise risk management strategy to be adopted for the entire project duration. The risk register will be useful in this respect.

5.5 CHANGE CONTROL STRATEGY

Please describe what procedures are to be put in place to manage changes in project requirement otherwise known as scope creep.

5.6 PROJECT ASSURANCE STRATEGY

Please summarise the mechanisms to be put in place to monitor the performance of the project, for example the schedule of internal and external reviews.

Appendix 1C: Example of a partnering charter

Insert Project Title

"Working together to build a strategic partnership and achieve our mutual goals"

Partnering Charter

(1) To be delivered within a collaborative working environment of 'no blame' culture based on trust and respect and which will always seek to deliver the best possible outcomes for the project.
(2) To be delivered on time, within budget, to a quality to meet the requirements of the client and the building's end users.
(3) The client, as employers, recognise that they have a very positive role to play in the Project Team in order to ensure that the contractors are able to deliver against the client's challenging expectations.
(4) To create the right platform and environment where teamwork plays an important role in success.
(5) Delivered through collaborative working and early contractor/sub-contractor involvement, striving to make each and every pound go further and deliver more.
(6) Most importantly for all involved with the project to enjoy the experience of planning, designing and constructing an iconic building within the heart of the campus.

Appendix 1D: Example of a contractor competency questionnaire

Contractor Competency Questionnaire

As the client, the (organisation removed for confidentiality) has a duty under the Health and Safety at Work etc. Act 1974 (HASAWA) and the Construction (Design and Management) Regulations 2015 (CDM) to ensure that the Contractor engaged for work is competent to meet the obligations conferred on them by law.

To ensure we can meet the requirements of this duty the (organisation removed for confidentiality) use the Safety Schemes in Procurement Registered Members Scheme (SSIP) as a basis for evaluating contractor competence. If you do not have a current listing in SSIP that relates to the types of work required by the project please complete this questionnaire and provide evidence in support of your submission.

Declaration	
I confirm that the information I have given in this form is a true and accurate statement of my organisation's Safety Management procedures.	
Name of person completing this form:	
Name of organisation:	
Position in organisation:	
Signature and Date:	
Section 1 – COMPANY INFORMATION	
Name of Organisation:	No. of Employees:
Address	
Contact Name and Position:	
Tel No.:	Email:
Insurance: Employer's Liability Insurance held: **Attach a copy of your policy**	Insurer:
	Policy No:
	Extent of Cover:
Insurance: Public Liability (3rd party) insurance held: **Attach a copy of your policy**	Insurer:
	Policy No:
	Extent of Cover:

Section 2 – HEALTH and SAFETY INFORMATION	
1	**Health and Safety Policy** *You are expected to have implemented an appropriate policy, regularly reviewed and signed off by the Managing Director or equivalent.* **Attach a signed current copy of your company's health and safety policy** (including statement of intent, organisation for health and safety and arrangements for H&S management within the organisation relevant to the nature and scale of the work).
2	**Competent Advice** *Your organisation and employees must have ready access to competent H&S advice. The advisor must be able to provide general, as well as construction related, H&S advice.* Please provide the contact details below: Name: Address: Tel No: Email: Provide details of their health and safety qualifications, experience and relevant training that enables them to undertake this responsibility:
3	**Risk Assessment leading to safe methods of work** *You should have procedures for carrying out risk assessments and for developing and implementing safe systems of work/method statements.* **Attach sample risk assessments and method statements undertaken for similar projects.**

Section 3 – TRAINING and INFORMATION	
4	**Training, Information and Instruction** *You should have training arrangements in place to ensure your employees have the skills and understanding necessary to discharge their duties as Contractors. You should also have a programme of refresher training, to keep employees updated on new developments and changes in legislation to good H&S practice.* How does your Company provide relevant health and safety training, information and instruction to its employees?

5	**Individual Qualifications and Experience**
	Employees are expected to have the appropriate qualifications and experience for the assigned tasks, unless they are under controlled and competent supervision continuously.
	Attach copies of qualifications or training certificates relevant to the assigned tasks e.g. Asbestos Awareness, CSCS, ECS, CPCS, SMSTS, SSSTS, CSIRS, PASMA, IPAF, S/NVQs, company-based training programme suitable for the work to be carried out, etc.
6	**Trades & Professional Bodies**
	Please provide details of any health and safety organisations, groups, associations or bodies that your company is a member of:
	Please provide details of any trade federations or other professional industry bodies or organisations that your company is a member of:

Section 5 – HEALTH and SAFETY PERFORMANCE

7	**Accident reporting**					
	You should have records of all RIDDOR reportable events for the last 3 years. You should also have a system for reviewing all incidents, and recording the action taken as a result.					
		Incidents reported to HSE under RDDOR				
	Year	Fatalities	Specified Injuries	Over 7-day injuries	Dangerous Occurrences	Non Reportable
	This Year (year to date)					
	Last Year (full calendar year)					
	Year before last					
	Enforcement Action					
	How do you investigate accidents and incidents?					

7	**Enforcement Action**
	You should record any enforcement action taken against your organisation over the last 5 years and the actions taken to remedy the relevant matters.
	Provide details of any enforcement notices issued or prosecutions taken against your organisation in the last 5 years and what action was taken to rectify matters, if none please state:

Section 6 – MONITORING, AUDIT and REVIEW	
8	**Monitoring of Procedures**
	You should have a system in place for monitoring of procedures, for auditing them periodically and for reviewing them on an ongoing basis.
	What methods are in place to ensure effective communication/consultation with staff, on health and safety matters, following a procedural review?
	How do you ensure health and safety procedures are followed by staff?
	Attach an example of a site inspection report from a similar project.
	Attach a copy of a site inspection report.
9	**Appointment of Subcontractors**
	You should have arrangements for appointing competent subcontractors/consultants.
	What criteria is used to assess the competence of any subcontractors?
10	**Monitoring Performance of Subcontractors**
	You should have arrangement in place for monitoring subcontractor performance.
	How do you monitor subcontractor performance?

Section 7 – REFERENCES	
11	*You should give details of relevant experience in the field of work for which you are applying.* Give the names and addresses of 2 organisations where you have completed similar work ideally within the last 12 months. |	| |---|---| | | | Provide details of an individual who has first hand knowledge of your involvement in the project below: Name: Address: Tel No: Email: Where there are significant shortfalls in your experience, or there are risks associated with the project that you have not managed previously, you should provide an explanation of how you would overcome these issues:

APPENDIX – Evidence to Support Submission	Attached (please tick)	
	Yes	No
Employer's Liability Insurance		
If No please state why?		
Public Liability (3rd party) Insurance		
If No please state why?		
Health and Safety Policy		
If No please state why?		
Sample Risk Assessments (for similar projects)		
If No please state why?		
Sample Method Statements (for similar projects)		
If No please state why?		
Health and Safety Training Records		
If No please state why?		
Qualifications and Training Certificates		
If No please state why?		
Example of actions taken following an investigation		
If No please state why?		
Example of a site inspection report (from similar project)		
If No please state why?		

Appendix 1E: Example of a health and safety contractor's handbook

CONTRACTOR HANDBOOK

> It is a condition of the employment by:
>
> The client wishes to make contractors or subcontractors, understand and take all necessary steps to ensure compliance with their Health & Safety obligations.
>
> In order to assist contractors and subcontractors to meet their obligations while working on the Estate, the Client has prepared Health and Safety Rules which are required to be read and complied with.
>
> A failure by any contractor to observe the provisions of rules may be viewed by the client's Appointed Officer as a potential breach of these rules. As a minimum, in the event of such a failure, the works will be suspended until the outcome of an initial investigation is known and conditions which are safe and without risk to health are provided. Furthermore, a formal written non-compliance will be produced and a copy registered on the contractor's file for future consideration.
> While every effort has been made to cover all important matters, it is not possible to give information covering every possible hazard. Should you wish for any further information or advice please do not hesitate to contact your client Appointed Officer.
>
> No work may take place on the Estate without written authorisation from your client Appointed Officer.
>
> The procedure for obtaining authorisation will be explained during the induction process.

HEALTH AND SAFETY RULES FOR CONTRACTORS

(1) CONTACT

The (client organisation) will appoint an Appointed Officer at the start of a Project.

The Contractor will be notified of the name of the Appointed Officer throughout the tender process and related correspondence.

(2) REPORTING ATTENDANCE ON SITE

You must initially report to the Estates Helpdesk, giving the name of your Appointed Officer.

The Appointed Officer will check if they have had an induction in the last 12 months and arrange if required.

Out of Hours Access – Following prior approval from your Appointed Officer, report your attendance by signing in and out of the register in Security.

(3) PARKING ARRANGEMENTS

Parking spaces are limited.

Parking will only be permitted in designated parking areas as agreed with your Appointed Officer.

Any unauthorised parking will result in a fine.

Vehicles must not cause any obstruction which would interfere with the normal working of the Company or access by Emergency Services.

Drivers are required to exercise due care and regard for the safety of others.

(4) FIRE ALARM AND EVACUATION

The fire alarms are tested weekly and will sound for no more than 20 seconds. The Appointed Officer will provide you with the relevant testing times.

Should the Fire Alarm sound for longer or at any other time vacate the building via the nearest exit and go to the Assembly Point which is displayed on notices throughout the buildings.

If you discover a fire you should activate the nearest fire call point, proceed immediately to the Assembly Point and contact Security from a mobile.

If you think you've activated the alarm due to your work activities notify the Building Controller without delay.

DO NOT return to the building until you have been informed it is safe to do so by the Building Controller.

(5) BASIC SAFETY RULES

- Smoking

    Smoking is not permitted in any building or near main entrances to buildings. Electronic Cigarettes can be used in offices/rooms which are not inhabited or on view to other staff or students; they must not be used in any circulation spaces/public areas.*

    *** Within a site that's been handed over to the Principal Contractor, their rules will apply.**

- Refreshment and Toilet Facilities

    If you are permitted to use the Client's refreshment and toilet facilities you should first change out of dirty work wear and keep the

Appendix 1E

facilities clean and tidy. For your own safety please adopt good hand hygiene practices; remember to wash your hands before preparing or eating food.
- Drugs and Alcohol
    No intoxicating liquor or drugs will be allowed on the premises except for Prescription Drugs.
    Where your site workers are required to take prescription drugs, it is your responsibility to ensure their work performance will not be adversely affected whilst on site.
- Conduct
    Ensure that your general behaviour and actions and that of your site workers does not cause any offence or disturbance to any member of the Client Community.
- Guarding
    Never operate with safeguards that have been altered, bypassed or removed.
- Noise
    The use of radios on campus is forbidden, any noise generating activities including drilling must be pre-arranged with your Appointed Officer to minimise disruption within occupied areas.
- Slips, Trips & Falls
    You must protect others in close proximity to your work area from falling objects, slips, trips and falls and any other risks you may create.
- Speed Limits
    The speed limit on the inner campus is strictly limited to 5 mph.
    The speed limit on Client Road is 10 mph and is subject to all regulations of the Highway Code.
- Housekeeping
    High standards of housekeeping will be maintained at all times and general work areas should be kept clean and free from obstructions.
    Adjacent areas to your work area must be cleaned regularly to avoid any buildup of dust.

(6) FIRST AID

To summon First Aid or the Emergency Services you should:
    ring from a mobile, stating the location and nature of the injury.
    This is the direct emergency number for the Security Office in Maxwell, please add this number to your mobile, just in case.

(7) REPORTING INCIDENTS

All near misses, accidents and incidents must be recorded and reported to the Appointed Officer as soon as reasonably practicable.
    This is in addition to any reporting you may do for your own organisation or to the HSE.

You must inform your Appointed Officer immediately if you:

- have, or discover a spillage;
- discover, disturb or damage suspected Asbestos Containing Material

    - STOP THE WORK, seal the area and report it without delay.

(8) RISK ASSESSMENTS AND WORK PERMITS

Where appropriate, site specific Risk Assessments and Method Statements (RAMS) must be provided for review by the Appointed Officer in advance of any planned work.

The responsibility for ensuring a safe method of work is adopted rests with the contractor.

Your Appointed Officer will inform you of the hazards; Asbestos, biological, chemical, electrical, mechanical, etc. in the area you are working* and any procedures you need to be aware of.

Asbestos registers for the areas where you are working will be made available by the Appointed Officer.

You must obtain Authorisation, in WRITING, before every project is started. This will be in the form of a 'Work Authorisation Certificate' or 'Permit to Work'.

At the end of the planned work the authorisation documents must be signed off and returned to the Appointed Officer.

To be issued with a Work Authorisation Certificate in addition to providing a Risk Assessment and Method Statement (RAMS) for review by the Appointed Officer, the person doing the work must be able to control the hazards, e.g.

- No isolations or line breaking
- No confined space entry
- No excavations
- No hot work
- Access/egress is reasonable

Where the risks can't be controlled via the Work Authorisation Certificate and RAMS alone a Permit to Work may also be required. A Permit to Work is a formalised document which authorises:

- certain people to carry out;
- specific work at a;
- specific site at a;
- certain time and sets out the main precautions needed to complete the job safely.

A hard copy of the permit to work should be clearly displayed at the work site for the duration of the works.

Permits must be signed off and returned to the Appointed Officer at the end of the work.

Where the work will be conducted in a site which has been handed over to the Principal Contractor then it will be their responsibility to arrange for adequate controls to be put in place.

Typically tasks which may need to be authorised by a Permit to Work will include;

- Asbestos Removal
- Confined Space Entry
- Excavations
- Higher risk electrical work
- Hot work
- Work on Fire Alarm Systems & Emergency Lighting
- Loft / Ceiling Void Access
- Roof access / Roof work

Contractors should confirm with the Appointed Officer whether or not a Permit to Work is required – this MUST be done prior to commencing the work.

(9) ELECTRICAL WORK

You must obtain a Permit to Work from your Appointed Officer prior to:

- Connecting to, or interfering with any electrical or other services.
- Entering any sub-station, switch room or similar area.
- Working on Live Electrical Systems is generally not permitted; except where it is necessary due to the nature of the work, e.g. testing, and a Safe System of Work must be in place.

Any work on electrical systems, however minor, may only be completed by a suitably trained and experienced electrician.

(10) ROOF WORK

If it can't be avoided; all work at height must have an appropriate risk assessment and method statement and if your work requires access to a roof your Appointed Officer will provide you with a copy of the relevant roof hazard sheet.

In addition, all roof access must have a roof access permit to work.

(11) FIRE SAFETY

All Client buildings are equipped with automatic fire detection systems.

If you are doing anything that might compromise the fire system, for example, generation of dust, blocking fire exits or extinguishers at least 10 days' notice is required to enable the Appointed Officer to make the arrangements to ensure unwanted alarms are prevented and necessary precautions are put in place.

Without exception:

- Flammable materials must be stored securely and appropriately and not left out unattended, particularly at night;
- Dust levels must be kept to a minimum;
- Don't allow combustible materials and debris to accumulate.
- Don't store materials and equipment on stairways and other escape routes.

(12) HAZARDOUS SUBSTANCES

All work which is carried out on Site must comply with COSHH Regulations:

- Chemical Substances:
    Prior to bringing chemicals into the Client, e.g. acids, oils, etc. you must provide the Appointed Officer with associated Risk and COSHH assessments for review, and they must be displayed at the site of the works for the duration of the works.
    Your COSHH assessments whilst based on any Safety Data Sheet provided by the manufacturers must also include how the materials will be handled, used, stored, transported and disposed of whilst on site.
- Dust:
    Where equipment is used which is known to generate dust, provision must be made by the contractor to contain the dust, and arrangements must be made to ensure the work is properly supervised.

(13) LABORATORY ACCESS

The Client has Biology, Chemistry and Physics Laboratories and numerous Engineering Workshops.
  Hazards in these areas may include:

- Harmful organisms
- Hazardous chemicals
- Lasers & Power tools

<u>If you need to access any of these areas:</u>
In addition to your Work Authorisation Certificate, the Appointed Officer will arrange for the Area Supervisor to provide a 'Certificate of Clearance' identifying any remaining hazards and associated controls.
  This will be displayed at the entrance to the room and you need to make sure you are aware of the residual hazards identified.
  If there's no certificate on the door or you find anything you are unsure of, STOP THE WORK and report your concerns to the Appointed Officer without delay.

(14) LONE WORKING

Anyone working alone must not be placed at any greater risk than any other employee working with others.

Where, following your lone worker risk assessment, there would be additional risk for someone working alone the Client will expect you to provide a second person to be in attendance at all times, so that if anything should happen they can provide or call for assistance.

This is particularly important where your team will be expected to work in high risk areas such as:

- working at height
- confined spaces
- laboratories
- electrical works and
- work close to exposed live conductors.

(15) LIFTING OPERATIONS AND EQUIPMENT

All work on site must comply with LOLER Regulations:

Any lifting equipment brought on to the premises must have a copy of its current inspection Certificate, for presentation to the Appointed Officer, before it is used.

Any person using a MEWP shall be adequately trained and hold a current certificate.

UoS cranes, hoists and lifting equipment must not be used by contractors.

(16) PERSONAL PROTECTIVE EQUIPMENT

Contractors must provide all appropriate PPE as indicated in the RAMS.

Protective equipment must be used at all times where necessary, regardless of your own views on risk.

Protective equipment must be worn in designated areas including labs.

(17) PLANT AND EQUIPMENT

Plant, tools, tackle and equipment brought onto site must be fit for purpose, tested, maintained and in good working condition.

Electrical equipment must comply with all current Electricity Regulations and must:

- NOT exceed 110V without prior permission from the Appointed Officer
- be 'PAT' tested

All machinery brought onto site must comply with the PUWER Regulations (Provision and Use of Work Equipment) and be guarded or fenced appropriately.

At the end of each day you must ensure all your equipment is fully isolated and locked away.

(18) VEHICLES ON CAMPUS

You must organise your work to allow pedestrians and vehicles to move without risks to health:

Traffic routes should be indicated by warning signs and barriers.

Delivery vehicles must not impede access for emergency vehicles and will only be allowed on site for the loading or unloading to be completed. Deliveries should be pre-planned with someone available to receive the goods or they will be turned away.

Reversing should be kept to a minimum but where required, reversing aids and banksmen should be used.

<u>Extreme caution should be taken whilst driving or operating machinery on campus due to the large numbers of students moving between buildings (particularly at the start and end of lectures). It is essential that pedestrians and vehicles are segregated.</u>

(19) WASTE MANAGEMENT

The Contractor is responsible for the removal of all waste from site in accordance with current environmental legislation.

Unless specifically authorised you must not place debris into skips controlled by the Client of Salford.

Skips left on site MUST be of the self-contained lockable type and their location agreed by the Appointed Officer.

Care must be taken not to discharge trade effluent or contaminated liquids into the drainage system or water courses, e.g. adequate storage facilities must be provided for diesel fuel to ensure containment and prevent spillage.

Detailed records must be kept of all waste removed from site including the type and volume of waste removed from site and the method of disposal (landfill or recycled).

(20) WORK AT HEIGHT

All work at height must comply with Work at Height Regulations; scaffolds, ladders and other access equipment must be in sound condition and of good construction, adequate for the purpose and properly maintained.

If as a result of a Risk Assessment ladders are identified as an appropriate control, as a minimum they must be:

- Used for access and egress only or;
- for work of short duration that is considered to be low risk.

Ladders must be inspected before use to ensure they are in a safe condition and they must be secured adequately before use.

Ladders must be taken down after use or at the end of the day.

Unattended ladders and ropes must be secured out of reach of students and other unauthorised persons.

UoS equipment, including ladders must not be used by Contractors.

(21) CONTRACTOR IN CONTROL

When a site is handed over to the Contractor (including the Principal Contractor), as a minimum we expect that:

A suitable site induction is provided to all construction site workers taking into account, but not limited to:

- the information included in this induction;
- any site specific risks and control measures that those working on the project need to know about;
- first aid arrangements;
- accident and incident reporting arrangements.

Necessary steps are taken to prevent access by unauthorised persons to the construction site including:

- Physically defining the site boundaries using suitable barriers and warning signs;
- Special consideration of the nature of the business is given (adjoining areas with student/staff access);
- Changing fence lines & access routes can only be carried out in agreement with the Appointed Officer.

Provision of suitable and sufficient welfare facilities.

The Client reserves the rights to carry out periodic site inspections to assess compliance with control measures.

FINAL NOTE:
It is your responsibility to ensure that all the information provided in this document, that is relevant to your works, is included in your risk assessments and method statements both for dealing with the issues raised as well as the work you will be doing. You will then need to ensure that all the site and safety information is passed on to your staff, subcontractors and anyone else that comes onto your site during the contract, through site inductions, tool box talks and any other means appropriate.

Appendix 1F: Example of a project execution plan (PEP)

SCHEDULE OF REVISIONS

Revision	Date of Revision	Details of Revision	Revised By

CONTENTS

SECTION A – PROJECT
    A1 – Project Summary and Key Information
    A2 – Regulatory Compliance

SECTION B – PEOPLE
    B1 – Project Directory
    B2 – Project Structure
    B3 – Consultant Schedules of Services
    B4 – Contractors' Appointments
    B5 – Direct Appointments

SECTION C – PROCESSES
    C1 – Communication
    C2 – Change Management
    C3 – Document Management
    C4 – Health and Safety
    C5 – VRM

APPENDIX A – HANDOVER CHECKLIST
APPENDIX B – RESPONSIBILITY MATRIX

SECTION A – PROJECT

A1 – PROJECT SUMMARY AND KEY INFORMATION

1 Project Name and Scope

1.1 PROJECT NAME

The Project is to be known as [insert name], and unless requested by the Project Board, this shall be the only title stated on all documentation including correspondence, meeting notes, documentation and drawing title blocks. Within the body of documents, and in informal communication, the abbreviation [insert abbreviation] may be used.

1.2 PROJECT BACKGROUND

Description of why the project is required, and explanation of project to date.

1.3 PROJECT SCOPE

Outline items to be included within the project.

2 Project Plan

2.1 PROJECT PLAN

Description of project timescales and the pressure on this.
The project has been established to achieve the following outline programme milestones:
  Please give elemental breakdown of programme

3 Project Budget

3.1 PROJECT BUDGET

Element	Original Cost Plan	Forecast Expenditure	Actual Expenditure	Variance

Element	Original Cost Plan	Forecast Expenditure	Actual Expenditure	Variance

3.2 PLEASE SEE COST PLAN IN APPENDIX D

4 The Site

4.1 LOCATION

Brief description on location of the building and why this location has been selected

Insert Picture of Location Here

5 Abbreviations

5.1 ABBREVIATIONS

School of Computing, Science and Engineering	CSE
Mechanical & Electrical	M & E
Project Manager	PM
Post Project Review	PPR
Quantity Surveyor	QS
Senior Management Team	SMT
Construction (Design & Management) Regulations	CDM
Principal Designer	PD
Change Request Form	CRF
Employers Agent (depending on Procurement)	EA
Higher Education	HE
Programme bar chart (duration, completion)	Project Plan

A2 – REGULATORY COMPLIANCE

1 Planning Status

1.1 PLANNING CONSENT

Give description on whether planning is required, the type of planning request and likely timescales.

2 Building Regulation Status

2.1 BUILDING REGULATIONS

Give description on whether Building Regulations approval is required, the type of planning request and likely timescales.

3 Legal Agreements / Landlord / Party Wall / Rights to Light Issues

3.1 LEGAL ISSUES

Highlight and potential or foreseen legal issues

SECTION B – PEOPLE

B1 – PROJECT DIRECTORY

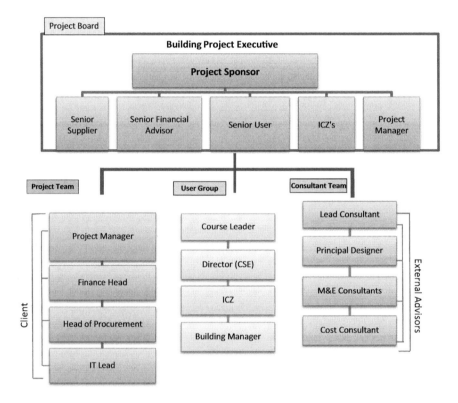

The Project directory will be maintained and updated by the Project Manager throughout the life of the project.

Role	Company Name / Address	Contact Name	Phone	Mobile	Email
Project Chair					
Senior User	"				
Senior Supplier	"				
Board Advisor / Head of Financial Accounts TBC	"				
Project Leader	"				
Project Manager	"				

Role	Company Name / Address	Contact Name	Phone	Mobile	Email
Construction/Design Team					
Main Contractor Construction Manager					
Lead Consultant Principal Designer Project Manager Lead Designer					

B2 – PROJECT STRUCTURE

1 Project Structure

1.1 THE COMBINED PROJECT BOARD/PROJECT TEAM

The Project Board will encompass the following roles with the following key responsibilities:

- Project Executive – Ultimately accountable for the project, key decision maker for project;
- Project Sponsor – Responsible for chairing the Project Board and championing the project to internal/external stakeholders.
- Senior User – responsible for clearly defining requirements and coordinating user interests;
- Senior Supplier – accountable for committing or acquiring the resources needed to satisfy the project; has authority to run the project within the constraints laid down by the Project Board.
- Project Assurance – independent monitor of all aspects of the project performance; and

The limits of the Senior Supplier and also the Project Board's delegated authority to make decisions are restricted to the works, budget and programme approved within the business case, and only when exceeded should an issue be referred to the Programme Board. The Project Board will be accountable for delivering the Project within the delegated authority, and will meet as a board on a as needed basis to:

- Accept the schedule of requirements and sign off the final design/plans.
- Respond to any exception reports (where a project cannot proceed within the delegated powers).
- Review and approve project continuation post tender.
- Accept handover / Completion sign-off.

2 Meeting Schedule

Initial Project Board Meeting	[Insert dates]
Pre-Start Construction Meeting	[Insert dates]
Mid Project Review Meeting	[Insert dates]
Handover and Close out Meeting	[Insert dates]

The list above is not exhaustive – if more meetings are required they will be appropriately arranged.

3 Public Relations

3.1 PROCEDURES

The Project Chair is responsible for all external communication relating to the Project, or an appointed designate. Should Project parties wish to communicate externally on project matters they should seek approval from the Project Executive, via the Project Manager before doing so. Should any incidents happen on site, or immediately adjacent, or with regard to the Masterplan development that may be of interest to the general public or news organisations, be it positive or negative, the Project Manager is to be notified immediately to allow the correct course of action to be agreed.

All Programme / Project team members are asked to promote good neighbourly and public relations. The Main Contractor will be required to sign up to, and promote the principles of the Considerate Constructors.

B3 – CONSULTANT SCHEDULES OF SERVICES

1 Consultant Services

The following consultants have been appointed as follows: (for contact details please see the Project Directory)

Client
Principal Designer
Lead Consultant
Quantity Surveyor

All design, specification and cost management services are to be complete by *(enter name)*.

2 Responsibility Matrix

Individual scopes of service are contained in the appointment documentation of the relevant consultant. The Responsibility Matrix in Appendix B below details the broad division of responsibilities on the Project.

B4 – CONTRACTORS' APPOINTMENTS

1 Main Contractor's Appointment

Please state who the main contractor is and how they have been appointed, include procurement method and evaluation.

B5 – DIRECT APPOINTMENTS

1 UoS Direct Appointments

Please name any direct appointments made by the client.

SECTION C – PROCESS

C1 – COMMUNICATION

1 Format / Pro Forma

1.1 CORRESPONDENCE/MEETINGS

In order for works to be carried out in a timely and cost effective fashion, it is important that all members of all project teams communicate effectively and efficiently. It is envisaged that the majority of issues arising throughout the Project will be able to be coordinated by the Project Manager.

Communication between parties involved in the project should predominantly be by oral (telephone and meetings) and electronic (email) means. It is of vital importance that a record is kept of all decisions made concerning the Project, and, therefore, an email (or letter) or minutes should be sent confirming the content of any decision made. Email communication should, where possible, only be used to issue information, a set of instructions or confirm a decision made, and should not be used in a 'conversational' manner to avoid inefficient practices.

Communication should follow the routes indicated and be between the key individuals from a relevant organisation. Those key individuals are then responsible for distributing that communication within their organisation.

Alternative communication routes may be followed in the event of an urgent situation or a party/parties not being available, provided that the course of action taken is in the best interest of the Project as a whole. In such a situation, the party/parties who would in the normal way have been involved, must be kept fully informed.

The Project Manager must be copied in on all relevant communications.

The hierarchy of correspondence is as follows:

- Minutes
- Email
- Conversation / Spoken

2 Meetings Schedule

Meetings form one of the prime opportunities for face to face communication and are considered fundamental to the successful execution of the project. Meetings will be organised on a needed basis. An agenda for a meeting should be issued with the minutes from the previous meeting. Generally minutes should be a succinct record of the key points discussed at the meeting, with an emphasis on recording the decisions taken and on future actions required.

3 *Correspondence Distribution*

All correspondence and documentation is to be issued with the Project title clearly referenced.

Minutes and other correspondence relating to meetings should be distributed to all meeting invitees, or as appropriate. Thought should be given to ensuring excess distribution is avoided; however, effort should be made to ensure all relevant parties receive information appropriately.

4 *Action / Information*

Each piece of correspondence should be clearly noted for each addressee as either action or information. This will assist all parties to prioritise issues and manage paperwork.

5 *Period for Reply*

All parties are obliged to respond to communications within the requested period. If an extension to this period is required, it must be agreed by all relevant parties.

6 *Reporting Arrangements*

All report problems should be addressed to the Project Manager in a timely manner.

7 *Contract Administration*

Drafting of the Contract and the administration of it is to be completed by ????????

C2 – CHANGE MANAGEMENT

1 *The process (to be reviewed subject to procurement process)*

The principle of change management is to establish an approved baseline of information, and if anyone involved in a Project suggests varying from it, a change would be raised for the team to consider, review and for the Senior Supplier to approve and instruct (if acceptable) or to escalate under an exception if it is outside of the scope of the delegated authority.

Unplanned change is highly disruptive and without the explicit control of change, there is a greater chance that project objectives will not be met.

Leadership from the client and active participation by the full project team are critical to the successful operation of a change management process.

The Change Management Procedure will apply to all change irrespective of the originator – e.g. Client, stakeholders, user groups, project team and other third parties.

The Change Management Procedure will apply to all change irrespective of scope, anticipated cost (subject to delegated authority levels).

The Change Management Procedure will be determined by the Employers Agent and agreed with the client prior to implementation.

When assessing changes, the project team may find the following classification helpful in determining which instructions are given priority:

- Corrective action – Errors or omissions in work already completed which will result in project objectives not being met.
- Request for change – A change which results in an alteration to the brief, design, specification or project acceptance criteria.

It is recommended that changes required for corrective action are prioritised so that agreed project objectives are met.

The Change Control System has two component parts:

- Stage 1 – Early warning notice issued and risk reduction meeting held.
- Stage 2 – Change Control. The formal process used to monitor change on a project. Change Control is described in section C2–3.0.

The Stage 2 Change Control Form is used to communicate a need for change and to analyse its impacts. **Up until approval the responsibility for the management of communication is held by the change originator.**

Use of the Stage 2 Change Control Form supports:

- Alerting all parties to all change – potential and actual, whether considered to have any implications or not.
- Circulating relevant information to all parties.
- Facilitating the assessment of the implications of change prior to formal client approval and instruction.
- Informing client decision making including identification of the actions needed to accommodate a change.
- Ensuring that the impact assessment and client approval does not delay the project.

- Managing change implementation to control the scope of the change so that consequences are within expected limits.
- Enabling the issue of a formal instruction to the project team once a change has been approved.
- Facilitating the central registration of all change.

The key duties and responsibilities in managing the change control system are as follows:

- Client – Sign-off of the change. Sign-off confirms that the client's recommendation is for the change to be implemented and acknowledges that the proposed change is the correct option.
- Originator – Driving the change assessment process prior to sign-off. The originator will complete and circulate the Stage 2 Form and will track and chase responses from the team to meet the agreed assessment timescale.
- Team members – Contribute to the assessment process as set out in Section 3.0. All members of the team have the opportunity to comment on all changes.
- Project Manager/Project Leader – Lead the assessment of completed Stage 2 Forms and manage the sign-off process.

In the event that a change is not identified by its originator prior to implementation, a Stage 1 Early Warning Form will be issued by the Cost or Project Manager to initiate the formal change process, **even if this is after the change has been incorporated into the design.**

The scope of change instructed must be limited to the work described on the Stage 2 Form. Whilst design development will be permitted following sign-off of the change; any material alteration to the scope of the change will require further approval via a further Stage 2 Form.

2 Stage 1 – early warning

Early warning is a key aspect of the change management system. The approach to early warning will be informal and the assessment requires minimal technical input. However, early warning must be used consistently if abortive work and possible delay and extra cost are to be avoided.

Early warning provides high level reporting of the cost implications of brief or design development at any stage of a project. It focuses exclusively on issues of affordability, which may include the cost implications of an extended development or build programme.

The Employers Agent will undertake a proactive role through the early warning system to identify change which is not identified by other members of the project team.

No comment on the broader implications of a change is required under the early warning system.

Use of the early warning system will not result in the issue of an approved Stage 2 change order.

The early warning system operates through the issue of a Stage 1 form.

Generally, Stage 1 forms will be issued by the Employers Agent as revised information is received. Other Project Team members are encouraged to request the generation of a Stage 1 form by the Employers Agent where they believe that an aspect of the project may be moving outside of its established budget.

3 Stage 2 – change control form

The change control procedure is the mechanism by which change orders are approved and issued for action. **No change can be issued without client approval obtained through the change control procedure**.

The change control procedure applies to all change proposed after Contracts have been executed.

Changes that fall within the scope of the procedure include:

- Changes to the brief.
- Changes to project acceptance criteria.
- Changes to design or specification.
- Changes to working method/sequencing that might have an effect on key project outcomes.
- Changes to previously instructed Variation Orders.

The change control procedure is the mechanism by which change orders are approved and issued for action. **No change can be issued without client approval obtained through the change control procedure**.

The change control procedure applies to all change proposed after Contracts have been executed.

Changes that fall within the scope of the procedure include:

- Changes to the brief.
- Changes to project acceptance criteria.
- Changes to design or specification.
- Changes to working method/sequencing that might have an effect on key project outcomes.
- Changes to previously instructed Variation Orders.

The impact of the change will be assessed against the following criteria.

- Project scope and quality
- Cost
- Programme
- Safety

- Sustainability/Environment
- Risk/Opportunity.

The impact of change will be assessed against current baselines for cost, programme and Specification as defined in the Contract.

The assessment of the impact of the change should include consideration of a 'do nothing' option.

A request for a Stage 2 Change Control form is to be raised by any party whenever it is considered that there is, or may be a change.

The Stage 2 form will be issued by and channelled through the Project Manager for each party to the change process.

4 Change control contacts

Organisation	Name	Email
Client		
Project Manager		
Principal Designer		

5 Change references

Change number suffixes are based on initials of each project party. Due to duplication of initials the following references are to be used on the project.

Organisation	Codings
TBC	
TBC	
TBC	
TBC	
TBC	
TBC	
TBC	
TBC	

6 Response times

Response times relating to priority codes on the Stage 2 forms are as follows:

Priority	Response time
Low	10 working days
Medium	5 working days
High	48 hours

7 Levels of delegated authority for issuing Stage 2 change instructions

Subject to any level of delegated authority being agreed by the client, all change instructions must be assessed using the Change Control procedure prior to issue.

C3 – DOCUMENT MANAGEMENT

1 Document control

Throughout the Construction Period the Project Manager will be responsible for coordination of formal submissions and responses between the consultant team and the Client.

All members of the Professional Team will maintain sufficient hard and soft copy filing systems to enable effective storage of information originating from that organisation.

Design Information control will be administered by the Contractor. The Project Manager will be responsible for ensuring that all Project Team members comply with the protocols established as part of the system.

C4 – HEALTH AND SAFETY

1 Health and safety

UoS require that the correct attention be given to all health and safety matters arising from the Project in accordance with the CDM 2015 Regulations. The client will appoint a Principal Designer who will take the lead in planning, managing, monitoring and coordinating health and safety during the project.

All Project Team members have a duty to raise any concerns that they have regarding health and safety on the project as soon as they become apparent either directly with the Contractor (when appointed) or the client.

C5 – VALUE AND RISK MANAGEMENT

1 Value management

All members of the team must adopt a value based approach. This approach involves employing lateral thinking and keeping an open mind to question why things are done in a certain way, and trying to identify more effective and efficient methods (particularly their tasks on the project).

2 Risk management

Risk management has been established at Project level, and encompasses the review of strategic / high level risk. Key Project risks are identified to ensure that they are dealt with in a timely and appropriate manner. The Risk Management Strategy for this Programme will differentiate between project risks and project issues as follows:

> Risk – A risk is an uncertainty, which, if it occurs, will have a negative impact on the project
>
> Issue – An issue is something which is certain to happen, or has already happened, which will have a negative impact on the project

The Risk Management Strategy is, first, to identify risks and issues which could negatively affect the chances of achieving the Project objectives; second, to determine the perceived relative importance of each risk identified in order to prioritise them with respect to each other; and, third, to discuss how each risk/issue can be mitigated and who should take responsibility in each case. It is vital that all parties actively participate in this process as risk is subjective, and certain risks may be more important to one party than to another. The outputs from this process will be included within the Risk Register, by the PM, following each meeting. A Risk Management Workshop may be held to expand on the above, if deemed necessary. The Risk Management process will review the value criteria established, and focus on the risks to delivering those objectives.

APPENDIX A

{The following provides a list of typical requirements at completion. The Client requirements and contract documents should be referred to in creating a project specific list.}

AAVT Temporary			
		By who?	By when?
1.	Practical Completion certificate issued.		
2.	Building Control Certificates of Completion received.		
	Building Control approval confirmed and certificate received.		
	Copy of original outline and detailed approvals issued to Client.		
	Statutory signage installed.		
	Fire-fighting appliances installed and ready for use.		

3.	Design team confirming that Contract Requirements have been met.		
	Consultants' confirmation that the Contractors' design and the works comply with the Contract Requirements:		
4.	All test certification has been signed by required parties and issued.		
	A copy of all certificates is included in the Building Manual. Test certificates to be included but not limited to:		
	Building Control Completion Certificate		
	17th Edition Electrical Test Certificate		
	Earth Test Certificate		
	Water distribution pipework pressure test certificate		
	Water by-laws compliance certificate		
	Certificate confirming chlorination of water distribution pipe work		
5.	Building Manual, including as-built drawings is complete, approved and issued in required format.		
	Approved As-Built drawings issued.		
6.	Final account agreed.		
	All contract instructions signed off and copies issued.		
7.	Condition surveys agreed between Contractor and Building Management.		
	Building accepted by Building Management and handover recorded.		
8.	All snags rectified and signed off. Procedure agreed for defects & outstanding works.		
	Agreed procedure and programme for the recording and rectification of works or defects outstanding or arising completion checklist.		
9.	Statutory and main services authorities satisfied.		
10.	Project Review carried out and attended by all.		

APPENDIX B

Responsibility matrix

Project Board/Project Team

The Project Board is ultimately responsible for the project, supported by the Senior User and Senior Supplier. The Board's role is to ensure that the project is focused throughout its life cycle on achieving its objectives and delivering a product that will achieve the forecast benefits.

Throughout the project, the Project Board 'owns' the Business Case.

SPECIFIC RESPONSIBILITIES

- Oversee the development of the Project Brief and Business Case.
- Ensure that there is a coherent project organisation structure and logical set of plans.
- Monitor and control the progress of the project at a strategic level, in particular reviewing the Business Case continually (for example, at each end stage assessment).
- Ensure that any proposed changes of scope, cost or timescale are checked against their possible effects on the Business Case.
- Ensure that risks are being tracked and mitigated as effectively as possible.
- Brief corporate or programme management about project progress.
- Organise and chair Project Board meetings.
- Approve the End Project Report and Lessons Learned Report and ensure that any outstanding Project Issues are documented and passed on to the appropriate body.
- Approve the sending of the project closure notification to corporate or programme management.
- Ensure that the benefits have been realised by holding a post-project review and forward the results of the review to the appropriate stakeholders.

The Project Board is responsible for the overall business assurance of the project – that is, that it remains on target to deliver products that will achieve the expected business benefits, and that the project will be completed within its agreed tolerances for budget and schedule.

Business assurance covers:

- Validation and monitoring of the Business Case against external events and against project progress.
- Keeping the project in line with customer strategies.
- Monitoring project finance on behalf of the customer.
- Monitoring the business risks.
- Monitoring any supplier and contractor payments.
- Monitoring changes to the Project Plan to see whether there is any impact on the needs of the business or the project Business Case.
- Assessing the impact of potential changes on the Business Case and Project Plan.
- Constraining user and supplier excesses.
- Informing the project team of any changes caused by a programme of which the project is part (this responsibility may be transferred if there is other programme representation on the project management team).

Senior User

The Senior User is responsible for specifying the needs of those who will use the final product(s), for user liaison with the project team and for monitoring that the solution will meet those needs within the constraints of the Business Case in terms of quality, functionality and ease of use.

The role represents the interest of all those who will use the final product(s) of the project, those for whom the product will achieve an objective or those who will use the product to deliver benefits. The Senior User role commits user resources and monitors products against requirements. This role may require more than one person to cover all the user interests. For the sake of effectiveness the role should not be split between too many people.

SPECIFIC RESPONSIBILITIES

- Ensure the desired outcome of the project is specified.
- Make sure that progress towards the outcome required by the users remains consistent from the user perspective.
- Promote and maintain focus on the desired project outcome.
- Ensure that any user resources required for the project are made available.
- Approve Product Descriptions for those products that act as inputs or outputs (interim or final) from the supplier function or will affect them directly.
- Ensure that the products are signed off once completed.
- Resolve user requirements and priority conflicts.
- Provide the user view on Follow-on-Action Recommendations.
- Brief and advise user management on all matters concerning the project.

The assurance responsibilities of the Senior User are to check that:

- Specification of the user's needs is accurate, complete and unambiguous.
- Development of the solution at all stages is monitored to ensure that it will meet the user's needs and is progressing towards that target.
- Impact of potential changes is evaluated from the user point of view.
- Risks to the users are frequently monitored.
- Quality checking of the product at all stages has the appropriate user representation.
- Quality control procedures are used correctly to ensure products meet user requirements.
- User liaison is functioning effectively.

Senior Supplier

The Senior Supplier represents the interests of those designing, developing, facilitating, procuring, implementing, and possibly operating and maintaining the project products. This role is accountable for the quality of the products delivered by the supplier(s). The Senior Supplier role must have the authority to commit or acquire supplier resources required.

SPECIFIC RESPONSIBILITIES

- Make sure that progress towards the outcome remains consistent from the supplier perspective.
- Promote and maintain focus on the desired project outcome from the point of view of supplier management.
- Ensure that the supplier resources required for the project are made available.
- Approve Product Descriptions for supplier products.
- Contribute supplier opinions on Project Board decisions on whether to implement recommendations on proposed changes.
- Resolve supplier requirements and priority conflicts.
- Arbitrate on, and ensure resolution of, any supplier priority or resource conflicts.
- Brief non-technical management on supplier aspects of the project.
- Attendance at monthly Joint Contractors meeting, the monthly Advice meeting, the Project Board and site visits.
- Sign off samples with support from the technical advisor.
- Line manage the Project Support Office.

The Senior Supplier is responsible for the specialist integrity of the project. The supplier assurance role responsibilities are to:

- Monitor potential changes and their impact on the correctness, completeness and integrity of products against their Product Description from a supplier perspective.
- Monitor any risks in the production aspects of the project.
- Ensure quality control procedures are used correctly, so that products adhere to requirements.

Project Manager

The Project Manager has the authority to run the project on a day to day basis on behalf of the Project Board within the constraints laid down by the board.

The Project Manager's prime responsibility is to ensure that the project produces the required products to the required standard of quality and within the specified constraints of time and cost. The Project Manager is also responsible for the project producing a result capable of achieving the benefits defined in the Business Case.

SPECIFIC RESPONSIBILITIES

- Direct and motivate the project team.

- Plan and monitor the project, and prepare and report to the Project Board through Highlight Reports and End Stage Reports.
- Take responsibility for overall progress and use of resources and initiate corrective action where necessary.
- Be responsible for change control and any required configuration management.
- Liaise with the Senior Supplier to assure the overall direction and integrity of the project.
- Identify and obtain any support and advice required for the management, planning and control of the project.
- Liaise with any suppliers or account managers.
- Ensure all identified risks are entered into the risk log.
- Day to day main point of contact with construction related activities.
- Attendance at Project Team monthly meeting.
- Work with Technical Advisor to agree sign-off for change/detailed design approval.
- Appointment of Technical Advisors, i.e. approval of professional fees.
- Working with the Employers Agent to ensure delivery and administration of the roles in accordance with the contract.
- Ensure coordination of the construction works with the operational activities of the campus.
- Monitoring and reporting to Project Board the overall project budget.

Stakeholder lead

SPECIFIC RESPONSIBILITIES

- Lead on stakeholder engagement with staff and students.
- Provide support to the Project Manager.
- As Head of Facilities ensure via Building Managers that the daily operations of the estates are not impacted by the ongoing construction works.

Project assurance

Assurance covers all interests of a project, including business, user and supplier.

SPECIFIC RESPONSIBILITIES

- Thorough liaison between the supplier and the customer is maintained throughout the project.
- User needs and expectations are being met or managed.
- Risks are being controlled.
- The Business Case is being adhered to.
- The right people are planned to be involved in quality checking at the correct points in the product's development.

- Staff are properly trained in the quality checking procedures.
- The right people are being involved in quality checking.
- The project remains viable.
- The scope of the project is not 'creeping upwards' unnoticed.
- Focus on the business need is maintained.
- Internal and external communications are working.
- Applicable standards are being used.
- Quality assurance standards are being adhered to.

Project administrative support

SPECIFIC RESPONSIBILITIES

The following is a suggested list of tasks:

- Administer change controls.
- Set up and maintain project files.
- Establish document control procedures.
- Collect actuals data and forecasts.
- Administer Project Board meetings.
- Assist with the compilation of reports.

Part 2
A construction risk management model for clients

Russell Whitaker

13 Executive summary briefing

(1) Part 2 introduces our Client Reference Guide and Construction Enterprise Risk Management (CERM) model. It is designed to complement the Royal Institute of British Architects (RIBA) Plan of Work and is written in the form of a Client's Job Book.

(2) The construction industry contributes to about 6 per cent of UK GDP. The private sector spends circa £24 billion on construction and over £4 billion of that amount is spent on the University sector, small offices and similar smaller projects, primarily smaller institutional clients.

(3) In every construction project there are at least three interdependent parties involved in a risk and reward venture; the Client, the Designer and the Contractor. It is the client that initiates the project, and procures the expertise required to design and build it, and it is the client that takes most of the risk.

(4) Risk in construction is particularly high for clients because, inevitably, architectural design for non-developer small and medium-sized enterprises (SMEs) will be a prototype. That is to say that the design team have to contextualise the client's raw idea into a form that it can be delivered within legislative constraints, the site of construction and the established methodologies and practice that are unique to the construction industry. Potentially this can be both a culture shock and a threat to the client's business model.

(5) As the client familiarises themselves with the design it may evolve thinking on the business model but at a stage in the design process when deadlines and the fixed timelines of due process leave little room for manoeuvre, creating a risk for the design team.

(6) Then there is the effect upon the business itself. A construction project starts and ends as a business change project. For many clients the design is not an end in itself but only episodic in a larger operational delivery programme. Many clients are simultaneously dealing with the new culture and protocols of construction at the same time as they are going through a process of change management within their own organisation. It is very often difficult to separate the two and the risk is that the design process itself becomes the tail that wags the business dog. In this book we talk about this as 'corporate construction risk' to capture the impact a construction project has upon the client organisation.

(7) This book does not seek to change any of the established facts or protocols within a client's normal approach to the construction industry. Instead it refocuses the process around the most important construction stakeholder – the client – and in a pragmatic way seeks to inform them of their responsibilities in construction and the established pattern of behaviour that regular and successful construction clients follow to de-risk a construction project.

(8) Construction literature habitually cast the client as a 'risk', when seen through the lens of designers and constructors without fully seeking to understand the client's perspective. We aim to help small corporate and institutional clients by bringing together the best of these processes and putting them into a single rational model, which we call the Construction Enterprise Risk Management (CERM) model.

(9) This model provides a framework of risk reducing measures and a plan of approach to construction from a client's point of view. Our model is not prescriptive, and any part can be left out depending upon individual risk appetite, however it is a rational sequence of events and parallels the RIBA Outline Plan Stage of Work widely used by the designers and constructors that make up the construction industry. Our model starts at least two years earlier than the RIBA Plan of Work, when most clients develop their initial thoughts.

(10) Chapters 14 and 15 explain typical construction clients, how corporate ideas become construction projects and the risks that entails for corporations themselves. Chapter 15 discusses the client's unique contribution to a project and how this is evolved into a risk management model that not only supports the design and construction process but also de-risks the client environment.

(11) Chapter 16 introduces the *Client-Side Construction Enterprise Risk Management (CERM) model* and the various component parts that form an integrated set of risk management procedures.

(12) To cover the extraordinary scope of the client's involvement in a construction project we can only be general in our guidance. We have therefore developed a set of supporting tools on our website which is complementary to subscribers of this book.

(13) This book also aims to start a debate about the professional construction client. In the UK designers and contractors have an established professional Body of Knowledge (BOK). There is no client equivalent and the establishment of a common client body of knowledge can only help clients, designers and contractors alike to reduce construction risk.

(14) In the field of risk there are the *Naïve* (that is, not recognising that one must go through a process of cultural change to deal with risk) to the *Novice* (that is, recognising that change is necessary without having the means to achieve it).

(15) This book is aimed at least to converting *Naïve* construction clients to become enquiring *Novices* and to become a network that shares experiences

and supports each other. We want to encourage readers to share their advice and include contact details at the back of this book.

(16) Finally, our process is deliberately designed to support the client's engagement with the protocols of the construction industry. It is complementary to the RIBA Outline Plan Stage of Work and our objectives are to create an explicit role for the client in a way that enhances the project as a whole, supports designers and mitigates project risk. This should only be welcomed by all sides of industry as a serious attempt to support an undersupported and essential construction stakeholder.

14 Construction clients, business propositions and corporate construction risk

14.1 The construction client

> A client can be defined as a party that is freely choosing to avail themselves of services and contracts with others for the supply of construction goods and services (Atkin and Flanagan, 1995). At the end of the relationship the client has ownership of the outcomes (Miller and Evje, 1998), with legal jurisdiction of the economic advantage (Hillebrandt, 1985).
>
> (Boyd and Chinyio, 2006, p. 5)

This chapter explores how client see themselves and what they are trying to achieve in a building project. It sets out the client's behaviours, how design teams respond to client behaviour and ultimately how risk is transferred between parties. It creates the notion of corporate construction risk.

How clients create and sustain profit making structures varies immensely and is beyond the scope of this book. However, one common factor is that the structure and hierarchy suits their core activity and gives them a naturalised stability. There will be differences between corporations (fundamentally between the decision making structures of public and private bodies) but all corporations have a *strategic, tactical and operational* function.

A construction project is a change project initiated for a business purpose. In constructing a building, we are seeking to change the way our organisation works. This presents challenges as we move away from our naturalised stability and creates risk to the corporation and the project itself as the client attempts to manage change within the organisation and the process of construction itself simultaneously.

This is further complicated by the fact that in construction sometimes we are in control and sometimes we feel like passive bystanders that are drawn inexorably into something that we did not plan for and that limits our options to manoeuvre. This is in effect a transfer gap. A gap in knowledge between what we had planned the project to be and how it evolves because of circumstances, and we need a way of measuring and controlling whether change is a healthy benefit to the business proposition or a costly distraction.

Successful small companies are naturally cautious and the disruption that construction creates can be likened to a 'managed major impact'. A bad construction project can be a major marketing disaster. However, a successful and well managed project can be evidence that the company can smoothly manage a major stage of a change process. The essence of project success is the company remains in control and the full scope of the impact and outcome is foreseen, planned and managed.

Our book is aimed at the infrequent constructor or the uninformed improver. It might be that our reader has undertaken a project and wants to approach the next project with a renewed sense of management and control. What we are trying to do is to take the innate dynamic capability within a corporate organisation to enable them establish a unique set of business skills – a map of what to expect in construction and a process framework that mitigates, what we call **corporate construction risk.** The aim is to help the average businessperson steer through construction making a unique and valuable contribution, from the front.

14.2 Construction clients and value propositions

What is the value proposition that makes it worthwhile for a company to take on the risk of a major project?

The decision to undertake construction is generally an incremental process established by utmost necessity and conviction that it is the only way forward.

The gestation period for project planning within a corporation can be, typically, up to one to two years before a decision to commit to construction is taken. The key client reasons for initiating a project, the underlying management tools and client drivers are as follows:

- **Sales or marketing driven** – Project that can be directly related to market impact/sales growth and a payback – possibly related to a brand launch.
    <u>Underlying management tool</u> will be cashflow/sales forecast – typically the baseline being year three of operation, assuming the first two years would be operating at a loss. Construction project will involve high street or customer impact driven re-location via land purchase or leasehold, or occasionally refurbishment of existing stock.
    <u>Client drivers</u> will be location sensitivity, market timed delivery, cashflow control and 'value driven' development – everything related to sales.
- **Cost Driven** – Project that saves duplication of administrative or other functions in the overhead by merger or co-location.
    <u>Underlying management tool</u> will be cashflow with emphasis of reduction of key corporate budget lines. Construction project interlinked with parallel processes, i.e. staff reduction, relocations and restructures.
    <u>Client drivers</u> will be speed of cost savings, tight management of costs and building in sense of new start or vision.
- **Strategic** – Project related to a vision, may involve improved sales or cost savings, or the individual budget head is accepted as loss making to provide greater strategic purpose or service, i.e. creche or other support facility.

>> *Underlying management tool* will be cashflow with emphasis of control of key corporate budget lines and marginal sales. Construction project interlinked with parallel processes, i.e. staff recruitment and marketing.
>> *Client drivers* will be cost control and strategic impact, cost efficiency to contain annual cost to within accepted margin.
> - **Whole Life Package Operation** – Project that is provided as a complete package service provision. Capital funding for premises is provided by third party and funded through annual charge to client.
>> *Underlying management tools* will be cashflow and containment of costs and overhead.
>> *Client drivers* are in reality the drivers of service operators rather than the corporation which is providing the space for the deal.

Some clients may see a new space as the only way of delivering value, but that decision has to be fundamentally driven by a rational value proposition, which may be improved by its relocation.

14.3 Client and construction team behaviours

14.3.1 Client behaviour in project development

The behaviour of clients approaching the construction market for the first time is most likely governed by the approach they take to most business ventures. Client objectives would be determined by the individual characteristics of the value proposition but primarily the main drivers would be:

- Is it the right product/process offer?
- How does the introduction of a new product/process improve the corporate position or make it more secure?
- How do we ensure financial sustainability and resilience to the corporation during and after delivery of the project?

This determines client responses to construction design and construction projects in two ways:

(1) _Design is a single aspect of a much longer process for clients_. It is likely that clients will find the BS6079 project cycle most closely aligned to their overall business project requirements for the whole life cycle asset management of a building.

   The construction industry widely uses the RIBA Outline Plan Stage of Work as its template for design and construction participation but in our opinion this is primarily focused around the feasibility and implementation phase and does not cover the whole life cycle of a typical BS6079 project.

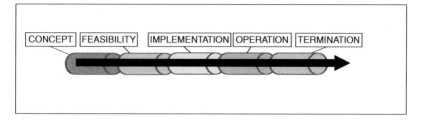

Figure 14.1 Project lifecycle BS6079

RIBA Stage 7 has tried to capture the 'afterlife' of a project but in our experience this needs fuller development to be fully adapted to the client's needs.

(2) Furthermore, to a client a design is the best first pass of an evolving concept *vision and business case* subject to *continuous review* and beneficial incremental variations are necessary at any stage to improve better business delivery. To an architect *the Strategic Brief* is the base layer to an incremental approach of staged design development under the Outline Plan of Work.

This means the starting point of each party is fundamentally different. Any client proposing to formulate their business case at RIBA Stage 0 is leaving it late and developing critical business fundamentals during Stage 0 prolongs this stage.

What many design teams fail to appreciate about client behaviour is that:

- To a corporation/client the project and its management starts and ends as a business venture.
- When we define business product or process, construction is very often an enabler to the business objectives of a client and is not an end in itself, but part of a greater whole (e.g. a BS6079 project lifecycle), in management and in risk terms.

14.3.2 Design team behaviour in design and construction

The design process is where the client's business outcomes are given a physical form. It is process of creative interpretation and for many design teams the objective is to start off with a coherent client business vision and provide an exciting design to fulfil it supported by trusted solutions.

14.3.3 Client behaviour in design and construction

Design and production are part of the normal routine of a client's world so why is building design and construction any different.

(1) Clients can be *construction naïve*:
> '90% of those who commission construction only do it once or twice. By monetary value 20% of construction clients commission 80% of the workload' (Ullathorne, 2015, p. 31).

In the view of many professionals:
> 'Clients ... need to be aware of what it takes to create a prototype that works perfectly, first time ... without crashing' (Ullathorne, 2015, p. 3).

(2) Construction can be a *catalytic process*:
> A construction project starts life as a project to support organisational change. However, the project itself can evolve unplanned organisational change as the possibilities within the new space become clearer. This can have unpredictable consequences on the original scope and cost of the project as the client varies the design late in the design development process.

(3) *Power and ownership:* given the definition of a client involves *ownership of the outcomes* paradoxically the nature of architectural design can limit the client's freedom in decision making giving them *conditional access* to design development. Even when design decisions may ultimately affect *the client's degree of economic advantage*.

14.3.4 The client behaviour and the RIBA Outline Plan Stage of Work

The client's needs in terms of business, user requirements, operation and maintenance requirements are established through the construction industry's standard client requirements processing system, the RIBA Outline Plan Stage of Work.

This is an incremental journey through the design stage, starting at Stage 0/1 (Strategic Definition, Design and Brief) and then developing the next layer of design and complexity, signed off by the client before the next stage proceeds. The design team establishes the 'voice of the client' (Griffin and Hauser, 1992), a systematic way of establishing and incorporating clients' wishes within a manufacturing process.

The risk of the solution based approach proffered by the industry is that it shifts the focus away from the requirements of the client to the needs of the designer. 'There is an inherent tendency for the client to be influenced by the preference of the designer' (Kamara et al., 1999, p. 320) effectively the 'design stage tail wagging the business dog'.

Furthermore, this 'layered' design approach, which at later stages is integrated with other disciplines can present a major problem for clients who want to have the freedom to change the design at any stage. If this results in significant design implications after stage sign-off, potentially further fee recovery can be pursued by the design team.

Understandably, the RIBA standard contract used to engage architects has many features designed to protect architectural income and reducing liability. Clients should be aware of the implications of the RIBA contract before engaging an architect.

The greatest risk to all parties in a project is an uninformed client's brief without boundaries or its own structure – that results in an insecure client's brief evolving in response to the developing design rather than leading it. Furthermore, this leads to an 'open' process, which fails to contain an ever expansive design process.

Clients like the idea of allowing the architectural model to shape their business ideas, but rarely are they prepared to pay for it. Clients should seek to manage design decisions and reconcile them with business objectives early on in the process.

Ultimately design, like any other business process, is a negotiation. Each party jostling their position to retain profit, reduce liability/risk and improve effect. The result should be a beneficial tension and 'push-pull' effect, however most often it can lead to the client feeling compromised, disempowered and as the instigator of the project and main risk taker feeling the 'economic advantage' has been compromised. This is neither good for the client nor the design team.

14.4 Risk management principles for clients in construction

Business clients are not unused to risk. Risk is prioritised in relation to objectives and is considered as a threat or an opportunity. In summary:

- Risk as a threat is defined as level of exposure which is considered tolerable.
- Risk as an opportunity is a measure of how much is put at threat to obtain the benefits of an opportunity.
- Risk appetite is the amount of risk that an organisation is prepared to be exposed to at any one time.
- Risk tolerance reflects what boundaries the company is prepared to allow on a day to day basis.

14.4.1 Corporate construction risk

How do corporate bodies gauge the nature of construction risk? To explain this, we have created the term '**corporate construction risk**'.

We believe **corporate construction risk** is generally classified into five major areas. Essentially the scale and size of this extraordinary commitment to a construction project will impact upon the following areas:

- *Strategic risks* – relating to the overall vision and outcome of the project and its impact upon existing organisation.
- *Financial risks* – relating to overall financial capacity and viability and to financial procedures and systems.
- *Legal risks* – those relating to liability and due diligence in carrying out the project.

- *Reputational risk* – the impact the process of carrying out the project has on the staff and public's perception of the organisation's ability to manage necessary complex change.
- *Operational risk* – the impact the project has upon the organisation's ability to maintain its current operation essentially understanding the project as a planned major impact and maintaining plans for business continuity.

Enterprise risk management is a framework and culture which enables the company to work within its risk tolerance and structure, or develop it dynamically in a measured and rational way. Our system is an enterprise risk management system for corporate construction risk(s).

Risk is increased at the moment a project starts to move the organisation beyond its normal operating boundaries and heightens as the client engages a designer and then a constructor.

Construction is a process of shared risks and rewards between client, designer and contractor. The common characteristic between all parties is related to the financial reality of doing business. The client wants the project completed on time and on budget. The designer and contractor both expect to meet profit and fee targets.

The additional factor for the client is *return on investment*. This accepts the inevitability of change and makes judgements on design approval decisions and variations on the cost (design and construction costs) against the expected benefit or outcome. Success for a client is not just about getting to the end of a project, it is also about the cost to the business of so doing and how close it came to the predicted result.

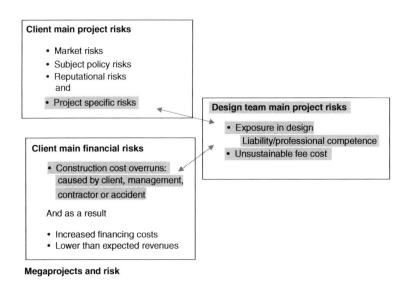

Figure 14.2 Risk dynamics in construction projects

Business risk is a generic and free-flowing concept designed to suit a wide range of business models. The UK Association of Corporate Risk Managers identifies five principles to achieve resilience:

- Ability to anticipate problems
- Adequate resources to respond to changing conditions
- Free flow of information to Board Level
- Capacity to respond quickly to an incident
- Willingness to learn from behaviour.

(Based on enterprise risk management (ERM) and ISO 31000, 2018)

As a contrast Project Management is much more systematic, is more plan-driven, with risks addressed stage by stage, and is based on MOR Management of Risk (2010).

Risks are workshopped, identified, assessed and controlled, a mitigation plan developed and a monetary value applied.

The approaches to managing risk within clients and design/construction teams are very different – the client process seeks maximum ability to manoeuvre whereas the design team process looks for 'locking down' solutions, stability and minimal change, in order to meet other client parameters of cost and time.

As a result, due to it being an uninitiated player in the construction world and its larger investment in the project the corporate client may have greater exposure to **corporate construction risk** without the standard protection mechanisms available to the regular players in the construction world.

14.4.2 Managing corporate construction risk

Most corporations believe they control project risk by *transferring* the risk to third parties (design teams) competent in the process. This is a fundamental error for three main reasons:

(1) The corporation initiates the project and process of change and can only manage the risk by evolving itself. The 'start-up' phase of a project is now widely seen as a critical risk stage for the whole project.
(2) Successful risk management is about corporation culture as much as it is about process, tools and systems. That is why this book focuses on cultural change which can adapt to any size of organisation or project (Enterprise Risk Management).
(3) Risk is essentially a process of managing challenges to the things we value and only the corporation can articulate the extent to which its fundamental values can change.

Corporate construction risk measures the impact of the design and construction phase of the project on the normal practices of the corporation (production, cashflow, etc.). The client is the initiator of the project but their first duty is to protect the existing corporate body.

The first step to managing risk is to understand the risk appetite of your organisation. The risk appetite will determine the measures the organisation will want to put in place (from our client-side process model) to mitigate risk.

To determine the risk appetite, we propose the Good Governance Institute's guidance on risk appetite (Good Governance Institute, n.d.) which contains their standard model for determining the appropriate response depending on project risk.

The second task is to determine whether you are a risk 'naïve' or 'novice'. According to a study undertaken on risk maturity by Wibowo and Taufik (2017) risk is best managed at organisational level not through just having protocols and tools at lower management levels. Being risk naïve means that you are unaware of the need for a change in organisational culture. Being a risk novice means you have recognised the need for culture change and are looking for guidance. The ensuing chapters set out the approach to a risk management system designed for novices.

'De-risking' a project is a fundamental pre-market activity. How the project is packaged, configured and presented to the market will ultimately determine the risk to the contractor and the cost to the client.

Decisions on delivery, risk transfer and impact on cost benefit can only be determined by the client, based on who is best equipped to manage that risk. Risk allocation takes place at the work package design/configuration stage, when the client is reviewing make (in-house resolution) or buy (outsourced) decisions.

Experienced business clients make these decisions every day within their own operations, yet continue to appoint a design team prior to making such a fundamental decision in construction.

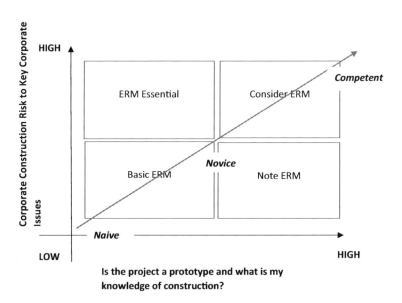

Figure 14.3 Risk status matrix and the use of ERM (CERM)

14.4.3 The professional clients' framework

There is no specific framework governing the client role. However, the seamless transition from business concept to design and build to asset management has parallels in a number of industries and can be selectively adapted for small/occasional construction clients as a guide.

If the client role can be identified, harnessed and understood as a recognised set of behaviours and skills that complement the RIBA and other design processes ultimately risk can be generally understood, properly allocated and reduced within project development.

The next chapter looks at the unique qualities a client can offer a construction project.

14.5 Summary: the need for a professional construction client

A successful project is a client responsibility, at every stage

The client is the main instigator of the project and the onus is on the client to establish, from the business case, the construction project objectives and the delivery mechanism. It is absolutely essential that the client ask themselves critical questions throughout project delivery, from initiation and at every stage approval:

- Can we afford it?
- Is it good use of money?
- What is the optimal return per unit of capital, is anything else better?
- What is the cost of debt and is this a valuable contribution against the cost of debt (a viable and valuable transfer of a capital to income generating asset)?

A successful project is a consequence of a successful business idea and how we manage risk

The first phase of a construction project for a client is the client genesis phase. This is a pure business development phase requiring at least one to two years before any design is initiated. The design stage and every stage after is subservient to techno-economic rationale. Corporate risk increases when clients diminish business clarity and vision in favour of a design which cannot evidence the essential features of business success – the process has failed and is the 'architectural tail wagging the business dog'.

A successful construction project is about designing the process to suit your advantage

'De-risking' a project is a fundamental pre-market activity based on packaging and configuration that enables clients understanding the design and construction market to reduce risk and secure the best price.

References

Boyd, D and Chinyio, E (2006). *Understanding the Construction Client*. Oxford: Blackwell Publishing Ltd.

Good Governance Institute (n.d.). Health and Social Care Integrated Joint Boards RISK APPETITE Aberdeen City Health and Social Carer Partnership. www.good-governence.org.uk.

Griffin, A and Hauser, JR (1992). Patterns of communication among marketing, engineering and manufacturing: a comparison between two new product teams. *Management Science*, 38(3), 360–373.

ISO 31000 (2018). Risk Management Guidelines: 03.100.01.

Kamara, JM, Anumba CJ and Hobbs, B (1999). From briefing to client requirements processing. In Hughes, W. (Ed.), *Procs 15th Annual ARCOM Conference*, 15–17 September 1999. Liverpool John Moores University, UK: Association of Researchers in Construction Management. 317–326.

MOR Management of Risk (2010). *Guidance for Practitioners*: ISBN 978011332740.

Ullathorne, P (2015). *Being an Effective Construction Client*. London: RIBA.

Wibowo, A and Taufik, J (2017). Developing a self-assessment model of risk management maturity for client organisations of public construction projects: Indonesian context. *Procedia Engineering*, 171, 274–281.

15 Unique client contributions to construction

15.1 Preamble

The previous chapter defined the construction client, behaviours, risk and reward in the current construction process. Its purpose was to argue how clients were most at risk in the process and how a more formalised proactive approach to their role was necessary to de-risk client-side issues within construction projects for the corporate body.

This chapter extends those key issues into a client-side management framework, what is known in the other industries as enterprise risk management (ERM).

15.2 The professional construction client

We know that all companies have a strategic, tactical and operational function and that constructing a building presents *a move away* from *the naturalised stability* of the operation.

Clients are also the main procurers of projects, with an impetus to initiate, amend and deliver the project mission (Winch, 2002) and who are in a significant position to influence the degree of supply chain integration and overall project deliverability (Briscoe *et al.*, 2004).

They are in prime position to 'de-risk' the project to themselves and to their stakeholders because of two fundamental truths:

(1) Project management has a systematic process of risk management focused on quantitative (measure based) risk assessment methods and 'supply-side' risk. Client risk is more fluid and interconnected with subjective and wider issues of corporate construction risk.
(2) The level of risk a consultant is prepared to take will be managed to within their own professional indemnity limits supported by a risk transfer mechanism (Professional Indemnity (PI) insurance), leaving clients to operate within their own assumed framework of risk management, within their own risk appetite and without a 'fall-back' mechanism.

Ultimately client behaviour has more of an effect on project outcomes (risks and rewards) driven by their perceived economic advantage (benefits capable of being measured).

15.2.1 Managing risk through dynamic change within the client organisation

The primary function of the executive should be to endorse the right degree of dynamic shift in the client environment to 'define project goals and execute in accordance with those goals'. This is a process of **sensing** (identifying and assessing opportunity), **seizing** (implementation and mobilising of resources to capture value) and **transforming** (configuring resources) leaving the client organisation temporarily transformed to deal with corporate construction risk.

Dynamic capability in an organisation is adopting a 'learned and stable pattern of collective activity through which the organisation systematically generates and modifies its operating routines in pursuit of improved effectiveness' (Zollo and Winter, 2002, p. 343) to adapt their current structure to the change that is taking place.

The degree to which the organisation will be adapted will be dependent upon their risk appetite. The lower the appetite for risk, the more of the organisation and our model they would wish to put in place.

Our Construction Risk Process Model is a universal response to the problem. As small and occasional construction clients we do not want to replicate the structures of a larger development organisation that permanently engage contractors, so by assessing risk as described in the previous chapter and using some or all of our adjustments to existing operations as described we should be able to reduce client risk associated with the project. This can be for the period of the project or a permanent reorganisation.

15.3 The four unique contributions a client can make to a construction project

We believe there are four unique contributions a client can make to a building project. These are inherent in their business skills but need re-focusing to construction.

- Clarity in purpose
- Realism and flexibility in operational planning and change
- Consistent client advocacy in design and production
- Balanced management of change and corporate risk throughout asset delivery and stabilisation.

These client behaviours and broad outputs form the basis of our ERM system for construction introduced in Chapter 16.

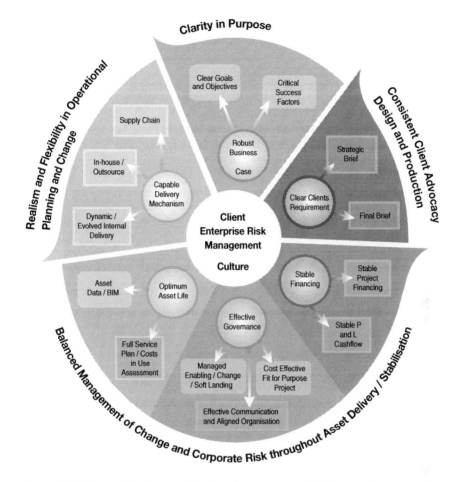

Figure 15.1 Client behaviours and the broad outputs of an ERM system for construction

15.3.1 Contribution 1: clarity in purpose: value propositions (outline business case, objective setting, cashflow optioneering, business case refinement)

In the last chapter we identified typical value propositions in construction. The value proposition for a construction project is the set of business beliefs that make the underlying project worth the risk. To make these useful to the project we need to develop these into a set of meaningful benchmarks (cost, quality, delivery, success) and objectives that provide our 'business red lines'. Corporations are naturally cautious and do not change a successful business model unless there is a good reason. To manage risk we need to know how far we are prepared to move away from the model and for what good reason.

In our model visioning and value proposition is the first critical stage of business concept development. Our 'business red lines' and key objectives and enabling objectives are established as part of the outline business case.

Design is about challenging given orthodoxies to enable improvement in business process. A construction project is by its very nature a catalyst for change initiated and controlled by the client. A corporation client must be clear which values will change (and how far) to enable successful business change. To manage this the client establishes clear objectives and a framework of delivery, including milestones, gateways and configuration.

The primary risk at this stage is establishing a budget without a design. We can however have a broad estimate derived from floor area and cost per m^2 to establish thresholds in borrowing, and business red lines to manage building cost within a range.

At the outline business case stage objectives are better carried out at least initially on an 'open protocol' basis (not subject to design restrictions) by a process manager, who would be a business specialist with some construction knowledge, rather than a project manager, who is more likely to focus on construction as a solution.

- Key business activities should be done by (internal professional) staff with no to minimal engagement of construction professionals at this stage until it is absolutely necessary. Internal business process staff should be trained in how to become professional design intermediaries.
- Clients are often surprised by high feasibility costs in design, when in reality much early development can be done by themselves.
- Effort expended prior to design team engagement is critical in reducing client project risk as well as ensuring a strong client voice and direction in the design engagement phase.
- Using a structured client methodology will result in a reduction in unnecessary professional fees and strong client led outputs and outcomes.

The alternatives are that we carry on as usual – that is, a design led solution.

15.3.2 Contribution 2: realism and flexibility in operational planning and change (work package agreement configuration, PESTLE analysis, stakeholder alignment)

The decision to undertake construction is an incremental process and the client business planning process (business gestation period) can be up to one to two years before a decision to develop is taken. The purpose of the operational planning phase of business gestation is to understand the total impact of the project upon the organisation and the implications that have to be managed by the organisation as a result of undertaking the project.

The output is a fully considered client side operational management plan.

In our model the objectives of this stage are:

(1) To make a difference through a 'holistic view' of the construction process and its impact upon the organisation – that is to say optimal performance and delivery for a joined up construction project delivered by the corporation.
(2) To provide corporate checks and balances so that the project can be delivered within the client's stated cost, time and quality objectives.
(3) To foresee and avoid exposing the corporation – and its business as usual function – to unnecessary risk.

In our model cashflow optioneering and management is a first stage activity as are all activities associated with preparing the organisation for the project.

Design and construction is a package managed alongside others, prepared and procured in a timely way. The focus at this stage is on limiting risk by careful package scoping and configuration.

The corporation should manage a project as a 'managed major impact' with use of cognitive mapping and other risk management techniques to develop likely risk scenarios. A 'de-risking' strategy should follow from this (prior to design team appointment) and undertaken as a fundamental client design activity.

How the project is packaged, configured and presented to the market will ultimately determine the risk to the designer/contractor and reduce cost to the client. Decisions on delivery and risk transfer and impact on cost benefit can only be determined by the client.

So often little time and adequate preparation is given to the operational planning stage and clients feel risk modelling is unnecessary. However, poor project preparation, undeveloped business ideas, limited operational planning often without a suitable client project environment and lack of adequate business impact planning prior to design and construction feature largely as 'top ten' reasons for project delay and cost overruns.

Furthermore, clients too readily assume they can transfer risk to a third party or insurer without knowing the cost of its full consequences or any evidence of the value benefit.

15.3.3 Contribution 3: A strong and consistent client voice in design and production stage (client requirements, design negotiation, develop design implications report)

Clients are generally categorised by the design and construction industry in respect of their knowledge of construction (naïve, partially informed, well informed) and the industry adjusts its expectations and rates to suit.

If a project is not a prototype it is likely the client will have an in-house team who share a knowledge of specialised requirements, They will also gravitate to the many specialist designers who have niche areas of design expertise for example in schools, airports and hospital architecture. The expectation that they and the client should agree a successful brief is high, because it is repetitive, and consequently corporate construction risk is low.

Genuine prototypes, however, are new ground, and changes create 'gaps and contradictions' (Briscoe et al., 2004). Neither the client nor design team know the outcome, yet the client's aspirations have to be reconciled for the project to be a success. Even the client's idea of what they want can respond and adapt to the change that is being initiated by the design. Corporate construction risk is high.

However, that is a transitionary process. Clients do not stay naïve and so called 'power asymmetry' (Kamara et al., 2002) – the power one party with specialist knowledge has in a relationship over another, is limited. Ultimately there is a rebalancing of dependency as the design moves on.

Clients should have three expectations from the design process:

(1) Client requirements will evolve as the design evolves.
(2) While the client has both *ownership of the outcomes* and *economic advantage*, the client requirements are only one of four main areas of RIBA Stage 0/1 briefing (the others being site requirements, regulatory requirements, and design and construction requirements) and there is a tendency for these other requirements to overshadow and diminish the precedence of the client's requirements. The client should always challenge the effect these have on business imperatives.
(3) Embarking on a design is often a catalyst for organisational change as well as scope change. This can have unpredictable consequences. This will need to be managed.

Also, clients are not passive in the design process:

- The construction phase has a bearing on the larger project envelope which wraps around it in which case the client has to be aware of the impact and reconfigure the project to meet expected outcomes.
- Clients may not know about design and construction but they know what the outcome should be of this phase of the project.
- Clients may employ specialist designers and contractors but will have critical and experiential thought on performance requirements.

In theory under normal rules of negotiation the client should have the right to revise the design and the team content if the original requirements and objectives are not being met. However, this 'creative process' in our opinion, is often perceived by clients as being compromised by the RIBA process and by client lack of preparedness for the process. Our system therefore aims to complement the RIBA approach, while developing an approach that suits a typical business format.

Our model starts with a clear business vision and objectives – this develops into the client requirements, a document prepared by the client before a design team is appointed.

The client has the right to choose the specialisms and change the team. If the design used during the design phase falls below expectation, then the client's only option may be to change the designer providing the solution – with good reason – as 'There is an inherent tendency for the clients to be influenced by the preference of the designer' (Kamara et al., 2017, p. 320).

A clear client requirements document enables options for design solutions to be generated and tested against business outcomes. In building design development solutions are often hidden, latent and unformed and successful reconciliation will only come not at a single planned point but through a series of interactions. This is the push-pull of developing a new design. The risk is that clients may pay a heavy price in late or bespoke design solutions or a delay in programme. The only solution would be to set aside difficult areas of design scope, seek another design solution and reintroduce them at a later stage. There are consequences and benefits, and the client should not feel compelled to accept a particular solution.

Our client process also assumes that design is a process of continuous negotiation and creative development during which the design team will seek to manage the 'voice of the client' and that the client will defend the 'red lined' value proposition outlined above. Both parties should plan and budget for the client's right to make reasonable changes as they merge business case and design into a single coordinated strategy.

15.3.4 Contribution 4: balanced management of change and corporate risk throughout asset delivery and stabilisation, introducing the Construction ERM (CERM) process model

The client is managing asset delivery yet there is no underlying framework for them to follow.

Our Client-Side Process Model sets out four key stages of delivery:

- Business Concept Development Stage
- Corporate Client/Delivery Capability/Transformation Stage
- Outcomes Delivery Stage
- Three Year In Stage

Our client delivery model in delivering the whole client side asset management lifecycle is closer to a full stage project lifecycle under BS6079 than RIBA Stages 0–7.

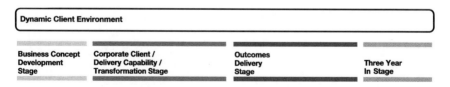

Figure 15.2 Client-side Construction ERM (CERM) model

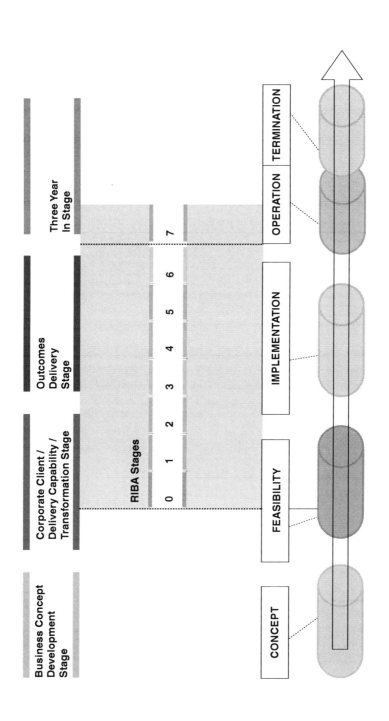

For clients the client process system is better mapped to BS6079 where assets have a fuller life cycle consideration including concept operation and termination stage omitted from the RIBA Plan Stage.

Figure 15.3 Relationship between CERM model against RIBA Outline Plan Stage and BS6079

Each stage of delivery in our Client-Side Construction ERM Process Model (Construction ERM Process Model) is an incremental step towards managing a successful outcome for the client before, during and after design and construction.

The process seeks to maximise the time to develop client-side activities before design procurement. Design is a fundamental activity but without context a disproportionate focus, time and energy of the organisation can be spent in this brief period in the delivery of the asset. Projects increase in budget and time for many reasons but corporate construction risk management activity as represented in our model seeks to reduce the likelihood and contain its impact.

Our model not only reduces risk for clients but the design team as well. One key driver of project success is client satisfaction. Success and satisfaction are not necessarily the same thing.

15.4 Project success and client satisfaction

Project success is a conflicting definition ranging from 'meeting technical standards' (Adam and Lindahl, 2017) to client satisfaction (Kamara et al., 2002), that is to say the client's feelings of pleasure or disappointment from comparing a product's *outcome* (in this respect the project initiation, design, procurement, construction and quality and cost) in relation to his or her expectations.

So, it is not just the outcome, it's the way we get there. Project success is also linked to the financial reality of doing business. The owner wants the project completed on time and on budget and the designer and contractor both expect to meet profit and fees.

15.5 Summary: the client as a unique contributor to project success

A successful project client will be prepared to change

They will understand they are a novice or naïve, understand their shortfalls and will go through a dynamic process of change to ensure project success.

A successful project client will have clarity of purpose

They will understand their vision and business model, but will recognise that design and construction is a negotiated settlement and they need a strong client voice in the design and construction process, supported by clear client 'red lines', essential for business success.

A successful project client will have realism and flexibility in operational planning

They will maximise their in-house business expertise and clarify their own operational plans, including aligning their organisation for change before engaging with designers.

A successful project client will have a strong and consistent client voice

They will understand their requirements may evolve but they understand that ultimately solutions may not present themselves immediately or with a particular designer or specialism. Their expertise and desire to actively ensure design success will evolve as time goes on.

A successful project client will have a systematic process of managing a business idea through design to asset delivery and post operational asset management

The RIBA Outline Plan Stage of Work is not designed to support the whole process of client involvement in a construction project nor the full range of activities required to reduce construction risk, therefore a new purpose built enterprise risk management model is necessary to ensure project success in the outcome and the process of arrival.

References

Adam, A and Lindahl, G (2017). Applying the dynamic capabilities framework in the case of a large public-sector client. *Construction Management and Economics*, 35(7), 420–431.

Briscoe, G, Dainty, A, Millett, S and Neale, R (2004). Client-led strategies for construction supply chain improvement. *Construction Management and Economics*, 22(2), 193–201.

BSI (2018). BS 6079 1:2010 Project management principles and guidelines for the management of projects.

Kamara, JM, Anumba CJ and Hobbs, B (1999). From briefing to client requirements processing. In Hughes, W. (Ed.), *Procs 15th Annual ARCOM Conference*, 15–17 September 1999. Liverpool John Moores University, UK: Association of Researchers in Construction Management. 317–326.

Kamara, JM, Anumba, CJ and Evbuomwan, NFO (2002). *Capturing Client Requirements in Construction Projects*. London: Thomas Telford Publishing.

Winch, GM (2002). *Managing Construction Projects*. Oxford: Blackwell.

Zollo, M and Winter, S (2002). Deliberate learning and the evolution of dynamic capabilities. *Organization Science*, 13(3), 339–351.

16 Reducing corporate risk using the construction risk management model

16.1 Risk, construction and clients

We have established that corporations are likely to have their business as usual risk working within an accepted risk management appetite, however they may have a high exposure to *corporate construction risk*. That is to say *the strategic, financial legal reputational and operational risks arising from a unique construction project having a perceived major financial and widespread operational impact upon the corporation itself.*

Previous construction experience is no guarantee to good risk management and corporations that have undertaken previous projects may yet wish to review their processes before embarking on further projects. A project may be successful in terms of its outcome but not in its implementation. Project success is as fundamentally about a managed and orderly execution of the project stages as it is about its outcome. Change is inevitable but were variations to the original plan properly analysed and discussed, were the implications fully known and was the corporation exposed to a higher risk than normal practice? This is as important as the outcome. This requires us to systemise processes that lead to risk and learn where the weaknesses are so we do not repeat mistakes.

In a construction project there are three interdependent parties involved in a risk and reward venture – the client, the designer and the contractor. It is the client that is the primary procurers of projects, with a key impetus to initiate, alter and deliver the project mission (Winch, 2010). Designers and contractors have established Professional Bodies of Knowledge such as the Royal Institute of Chartered Architects (RIBA) and the Chartered Institute of Building (CIOB), that manage and control the process of design and construction.

The small construction client has no body of knowledge, supporting their concerns, independent of the industry, so clients can only control risk by leveraging the advantage and the client voice within the constraints of established industry protocols and using the knowledge in this book/our website.

Risk is a dynamic process, which can be created, move and evolve through each stage of the project.

- At the <u>pre-project stage</u> the client manages risk within the parameters established by the initial raw business model and their own in-house expertise.

- At the design stage the raw business model will be refined as it is reformed into a design for a building by third party designers through strategic brief to final brief stage and new factors (extraneous to the client requirements, such as site and regulatory issues) are introduced that we can either contain within our established parameters or for which we must initiate new boundaries.
- At the design and construction rollout stage a further party is involved (the contractor) which may introduce more risks and may require an adjustment of boundaries. How far we have to adjust the boundaries at each stage depends on where we re-set them. That is always a process of acquiring knowledge, pragmatism in response to change and realism about cost and time.
- At the post-project handover to Three Year In Stage the risk is that the post-design output does not meet the benefits described at the pre-development stage. It is important that benefits and assets are constantly reviewed during the design and construction stages.

16.2 Client contributions to risk reduction

The client can reduce risk at each project stage of the construction process:

At pre-project stage

- Create a clear vision and compelling business case – communicate a picture of what is possible with a timeframe and steps to get there – get consensus and alignment behind a resourced plan for delivery.
- Create 'high level' configuration (client-side operational plan) for delivering the vision – each component part, its relationship with other parts, and a stage by stage mechanism for assured delivery within time, quality and cost parameters.
- Create/prepare organisation for project–client project environment – client portal – advise impact on 'business as usual'.

During project design and rollout

- Monitor client's interest through pre-design, design, delivery and post-delivery stage.
- Troubleshoot – use best solution to resolve corporate and project roadblocks to keep it on track/budget.
- 'Flex' the output, benefits and overall project configuration if necessary, to maintain success.

Post project

- Measuring/evidence success in delivery – against agreed benefits –understand lessons learned and re-employ them in continuous improvement.

- Update/maintain configuration records. Update client portal. Ready for post-building phase.
- Adapt the product – maintain fitness for purpose and high operational level and value asset throughout design life.

Our client-side construction model captures and formalises this contribution (see next section).

16.3 Navigating around the client-side Construction Enterprise Risk Management (CERM) model

The CERM model has two main levels:

- Strategic (framework) and
- Corporate management (activities and outputs).

and four stages:

- Business Concept Development Stage
- Corporate Client Delivery Capability/Transformation Stage
- Outcomes Delivery Stage and
- Three Year In Stage.

Within this framework there are a number of outputs.

The following chapters explain the outputs (products in project management jargon) that make up the client-side Construction Enterprise Risk Management model.

In our experience most corporations would normally produce these outputs in preparing for a project, so they are tried and tested. The unique proposition of our book is that the products are related in a rational sequence of progressive risk control measures. Our model provides a coherent system of enterprise risk management for construction clients that relate to the model used by the construction industry, the RIBA Outline Plan Stage of Work.

In this book and in our model we have attempted to provide construction clients with their own body of knowledge (BoK) as their counterparts in the construction profession that make up the industry have had for many years.

This cannot explain the products in detail, it can only provide an overview, therefore back-up material is contained in our supported website, which aims to be a source of client guides and experiences, to which we hope you will subscribe and through which we can continue to develop our model and give construction clients a body of knowledge.

This section describes what is to be produced at each stage and how it contributes to successful client managed construction projects.

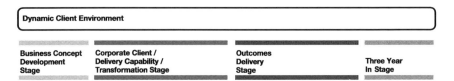

Figure 16.1 Client-side CERM model

- Chapter 16a discusses the Business Concept Development Stage.
- Chapter 16b discusses the Corporate Client/Delivery Capability/ Transformation Stage.
- Chapter 16c discusses the Outcomes Delivery Stage.
- Chapter 16d discusses the handover to Three Year In Stage.

Note: Three Year In Stage – the usual premise is it takes three years to establish the normalised (baseline) operating routine and costs of a business after change.

Each stage has a number of outputs, which make up our coordinated risk control system.

The output objectives and contents are described in Chapters 16a–16d in this book and supported by examples and case studies on our website.

Each chapter starts with a stage tracker. Each chapter describes the tools identified in the client process system in broad detail. More help is available with templates on our website.

Overlaying each stage is the dynamic client environment. That is the client expertise and systems that are necessary to reduce risk.

Each stage is constructed in the form of a job book. This is a set of tasks required to successfully complete that client stage. An easy reference guide to get you to the right part of the text:

Symbol	Purpose
🔔	The overall task in simple terms
📄	The product referred to in our CERM model
✍	Key headlines in the text of the relevant CERM product
👍	Notes and supportive text

Strategic direction is defined as two parallel lines:

Strategic Decisions: Sign off Vision, Value Proposition

Corporate management activities are defined as two parallel lines with a background:

Corporate Management Activity: *produce vision and value proposition*

16.4 Summary: managing risk using our client-side risk management model

Our risk management model provides a system for client risk mitigation

Projects may be successful in their outcome but poor in their implementation. If we know where mistakes have been made we can improve practice in the future.

Risk starts with the client, is dynamic and can move from pre-construction to construction and to handover. The client is a major player at each stage of project development

The client is the only project player who has to manage risk at all stages, and has clear contributions at pre-project, during design and construction and post-project stages.

Our risk management model provides a basis for a rational corporate risk culture

Clients initiate projects and therefore proactively have to establish key processes to mitigate risk based on the unique contribution that clients can make to a project. They cannot rely on a design and construct team to mitigate corporate construction risks that lie outside their scope but that may be a causal factor of project delay, increase in cost or scope, yet they can use our model to manage corporate construction risk through developing a strong enterprise risk management culture and organisational processes.

16a Business Concept Development Stage

16a.1 Purpose

<u>Business Concept Development Stage</u>: To develop an idea into a vision with value propositions/argument for change and steps and safeguards to initiate start-up capital.

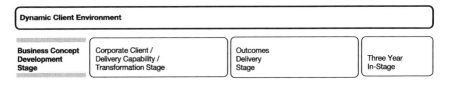

Figure 16a.1 Business Concept Development Stage

16a.2 Organisations and planning

No one organisation is typical but all companies share macro and underlying micro business planning activities and a business cycle all agreed at executive level:

Macro activities

- Business strategy vision and mission
- Business operation and configuration
- Business cashflow, finance management and risk profile.

Micro activities

- Business case development/market intelligence
- Operational rollout and structure (including design)
- Resource.

Business cycle activities

We have assumed a standard business planning cycle of:

- Review of strategy and business planning (typically September each year)/ 'calling in' of strategies/new business planning year requirements.
- Primary screening by executive.
- Strategic framework/business modelling and setting departmental planning/ budgets and objectives in the context of organisational delivery largely delivering the same product/processes.

  - Operational planning/roll-out
  - Cashflow iterations
  - Risk setting

- Sign-off by executive based on known/planned income and costs.
- Final agreement to next year and budget setting.

This can probably be described as a *business as usual* framework.

16a.3 Planned change and managed impact

Most organisations go through change processes to improve business productivity; however, these are usually incremental modifications controlled within boundaries of known and containable strategic, operational and cashflow limitations and they result in permanent change.

16a.4 Why construction change is different

A construction project is a risk because it is a prototype of significant size and budget; its skill set and responses are generally outside the organisation's competence. As the project's shape evolves with design, project boundaries that are not established, managed and controlled can rapidly cause a sudden and stressful intervention on the organisation's planned strategy, operational and cashflow processes.

In our experience the root cause is through poor project drafting by the corporation and through uncontrolled/ill conceived change beyond the original scope. We feel the impact can be mitigated through a systematic process of client development such as our CERM model.

When undertaking a project, it's a safe assumption that change is going to happen – so we should prepare for it and model the possibility and outcome of major risks through established models such as Monte Carlo analysis and Cognitive Mapping, much as the finance industry has done since the recent crisis in the sector.

The objective of the Business Concept Development Stage is to try and be as clear as possible about the vision and the value proposition and the risk at high level then drill down into the detail at Corporate Client/Delivery Capability/ Transformation Stage to manage delivery and contain corporate risk. Plan for at least one year for this stage.

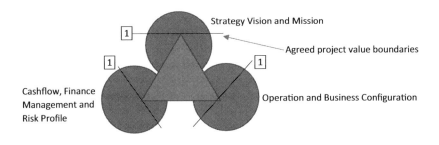

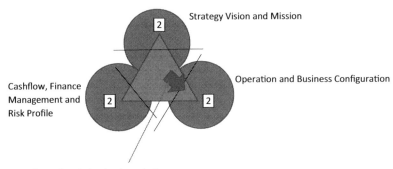

Agreed project value boundaries are being exceeded on one or all boundaries (position 2) going beyond the planned change agreed that allow corresponding organisational mechanisms to cope. Most likely these are irreversible and desired. The issue is that they are unmanaged.

Figure 16a.2 Change and business boundaries

16a.5 Business Concept Development Stage outputs

The **Business Concept Development Stage** includes three 'demand side' and two responding 'supply side' corporate management outputs. Together they form the outline business case. The general purpose and process is discussed below.

Corporate vision (demand) outputs describe the change:

(1) Vision and value proposition statement includes scope and exclusions.
(2) Funding strategy and cashflow optioneering statement (including sensitivity analysis and scenario planning).
(3) Risk probability assessment, transfer and assurance model.

Corporate construction risk mitigation (supply) outputs describe the process of:

(4) Client capability statement and de-risk strategy.
(5) Communication and stakeholder alignment strategy.

The significant factors in risk management are cultural – a willingness to use models to understand the problem, included in our CERM model as *supporting activities* and as the examples and toolkits provided on our website given at the back of this book.

16a.6 Job book: Business Concept Development Stage

16a.6.1 Corporate vision outputs

Strategic decisions: sign off vision, value proposition, create corporate governance, sign off corporate management outputs below

Corporate management activity: *produce vision and value proposition statement, funding strategy and cashflow optioneering statement (incl. sensitivity analysis) and risk probability assessment (incl. transfer and assurance)*

Briefing: What's the change, how do we manage cost, what does it take to deliver?

Vision and value proposition statement

An achievable aspiration for the future and broad timeframe – the key themes and broad impact to change in the three macro aspects:

- <u>Strategic change context:</u> change context (this might be supported by a political, economic, social, technological, legal and environmental (PESTLE) analysis) vision/mission, impact on operation and configuration and delivery, phases of change.
- <u>Objectives (high level configuration):</u> stepping stones to the vision and goals. The milestones underlying the objectives will form the configuration and work packages and our later delivery programme. There are two types:

  - *Operational objectives* relate to the delivery of the operation.
  - *Enabling objectives* relate to activities supporting the operational objective.

Typical project objectives are:

- To achieve a shared vision.
- To achieve the benefits set out in the business case.
- To develop the most cost effective and expedient purpose and method of delivery.
- To optimise the inherent capability within the organisation.
- To procure a capable and reliable supply chain to deliver the project.
- To create an accountable governing/operational management structure that:

- manages corporation risk
- communicates and receives progress and ideas
- is accountable for the product and its benefits.

Part of the later configuration planning will be to appoint the appropriate expertise to discharge the client requirements within and outside the organisation. This is generally based on skill set and capacity.

- High level costs: What are the costs of all the workstreams?
- Critical success factors (including construction): What has to be in place for the change vision/value proposition to happen and how will these deliver the benefits plan of the project? This provides the parameters for the client-side operational management plan in the next section.
- Benefits case: in strategic and financial terms what will arise from the change?

Funding strategy and cashflow optioneering statement

Finance strategy is broken down into three sections:

- Funding strategy
- Cashflow optioneering
- Client budget management control

Funding strategy

Most clients are used to a 'business as usual' corporate budget established from known processes. The project budget is new, unclear and without any scope or financial shape at the early stage.

At the Outline Business Case Stage, we will not have a detailed design (maybe an outline cost based on floor area) and we can therefore only present our funding strategy.

The finance strategy will depend on the business case and the project. Typical funding strategies are:

- Borrow and build
  This is the most common approach where the institution borrows the money and builds itself. The third party lender will require security in the form of a mortgage on the building; a payback period will be calculated which would normally be 25 years for a bank.
- Private equity
  The institution raises capital through the release of shares or private loans and this is recorded as a creditor in the accounts.

- Bond market
  The institution agrees to a loan longer than the 25 year bank limit, usually from a pension fund and usually with some form of share of income. This is generally pursued to provide lower interest payments and 'financial headroom'.

The key determinant is the risk appetite, sustainability and the 'financial headroom' required by the institution.

Cashflow optioneering

- At this stage when the scheme has reached feasibility stage a first broad estimate of the budget will be known. At this stage a number of variations of the cashflow will be prepared.
- The cashflow will be based on income assumptions and corresponding capital expenditure producing a cashflow baseline and then a number of variations based on likely scenarios and forecasting.
- The project will have a cashflow which will give a sense of whether the project will run to a surplus or subsidy after it reaches its full operating capacity (Three Year In Stage).
- At this stage any cashflow is a high level subsidy and surplus analysis at current values including soft delivery (operational set-up), enabling and capital/aftercare costs and other strategic considerations against expected (superior) benefits.
- Most corporations would undertake financial modelling based on an estimate of the added cost of the project over the forecasted (say) five year profit and loss (P and L) account.
- A number of scenarios may be drafted and in each case a capital expenditure risk analysis undertaken. Agreeing a working 'model' cashflow with known and unknown parameters is essential at this stage and this will have a bearing on the pacing and timeline of the project and ultimately the configuration.
- The speed of construction delivery will be determined by affordability as well as the corporate requirement for 'pace of change' and the date of the completed facilities' first trading day.
- Construction project environments are highly dynamic especially in the design stages and we have proposed an 'inner and outer circle' model of budget control in the next section. Some guidelines to bear in mind when producing cashflows:
  - Initial outline budget planning (early feasibility) high risk margins should be applied – up to +25 per cent error over design and construction costs.
  - Post stage 3/4 when design is more developed close to pre-tender estimates at which stage a more moderate risk margin of 5–7 per cent error over construction costs only can be applied – construction should be inflated to the midpoint of the tender. The Office for National Statistics (ONS) produces

a more robust figure for actual costs and may provide some leverage for reducing prices.
- Construction costs will be adjusted to a market tested rate at post tender.

Risk probability assessment, transfer and assurance model

Risks of construction on the organisation

We have previously identified this as *corporate construction risk*, which is the impact of the following caused by a construction project.

- Strategic risks
- Financial risks
- Legal risks
- Reputational risk
- Operational risk

It is widely recognised that it is the client who has the sole responsibility to:

- Define the project goals and execute them through the framework of a project.
- Initiate, alter and deliver the project mission and influence the degree of supply chain integration and overall deliverability of the project.
- Maintain the stability of the corporation during construction through managing cashflow and construction impact, that is to say the corporate construction risk.

Most corporate construction planning is only seriously initiated when a project professional such as a project manager is appointed. This is a fundamental error that ignores the inherent potential within client organisations and unrealistically places the onus on the design process to resolve unresolved pre-design stage risks.

Strategic decisions: sign off client capability statement, de-risk strategy and communications and stakeholder alignment strategy

Corporate management activity: *produce client capability statement and de-risk strategy part 1: reducing risk by having effective governance and a capable sponsor*

Client capability statement and de-risk strategy part 1

This sets out the framework in the broadest possible terms to deliver the vision. It describes controlled change processes, skills and resources required to achieve change and mitigate corporate construction risk by creating an enterprise risk management culture.

Dynamic Client Environment			
Business Concept Development Stage	Corporate Client / Delivery Capability / Transformation Stage	Outcomes Delivery Stage	Three Year In Stage

Figure 16a.3 Business Concept Development Stage

The first step to establish an enterprise risk management culture is to put the key risk management posts in place at strategic (board) and operational management level. In construction enterprise risk management this is a competent project sponsor/intelligent client and a proactive project board with clear objectives at pre-project, mid-project and post-project stage.

The key stakeholders, owners of the corporation construction risk and initiators of construction ERM and cultural change are:

- **Corporate governors** (*'strategic direction' in our model*) – the corporate board and its devolved project – board responsible for overseeing and resourcing the business case and managing the corporation construction risk.
- **Project sponsor/intelligent client**[1] (*corporate management in our model*).

The responsibilities of strategic framework and corporate management activity level are:

- Business case development/auditing
- Funding continuity
- Client requirements development and effective interface management
- Optimising performance
- Effective governance
- Organisational impact management
- Corporate risk reduction
- Handover management/in use management/asset management.

In our view the personal attributes of a project sponsor/intelligent client role would be a person that has:

- **Realistic/continuous time commitment** to project/access and influence at senior executive level.
- **General business knowledge** with **project skills.** Effective communicator at senior executive/project levels.

- **Empathy and commitment** to organisation.
- **Ability to rationalise problems in overall business/project context** before jumping to a solution.
- **Driven by continuous business justification, focused on vision,** able to evolve the project <u>and</u> business case to respond to changing circumstances.

Corporate management activity: *produce client capability statement and de-risk strategy part 2: reducing risk by sharing risk with a competent external supply chain*

Key client risks can be managed by:

(1) <u>Sharing risks</u> – continually keeping the risk owner under review – *risk is best located to the party best equipped to manage it* – and looking at the benefit of risk transfer at any stage of the project – even if it means a contract change is preferable to failing to manage the risk.
(2) <u>External supply chain management processes:</u> ensuring the right supply chain is appointed and key issues are identified early, and <u>interface management issues</u> identified.

<u>Corporate control, structure:</u> That is to say the corporate governors (strategic direction in our model) and the underlying project sponsor/intelligent client (corporate management in our model) structures.

Strategic decisions: sign off produce communication and stakeholder alignment strategy

---

Corporate management activity: produce communication and stakeholder alignment strategy

Communication and stakeholder alignment strategy

This an approach to managing any group affected by the project and making sure they understand the project's goals and impacts; their views are taken into account and their support engendered. This may involve stakeholder mapping, and determining from their relative power and influence the methods and frequency of communication.

Supporting activities

Enterprise risk management allows 'reasonable risk taking'. We have no certainty of events taking place so we have to model possibilities using an early risk workshop, probability and impact modelling and use of:

- Tools that help us to rationalise or predict the outcome root cause: Cognitive Analysis as a tool and Ishikawa Diagram (Cause and Effect).
- Tools that help us to predict the likelihood of one or another event happening: Probability and Impact matrices, Expected Monetary Value analysis, Monte Carlo analysis and Decision Tree.
- Having determined the risk and probability there are five methods to risk mitigation, namely Accept, Avoid, Transfer, Mitigate or Exploit.

16a.7 Summary: Business Concept Development Stage

The purpose of the Business Concept Development Stage

The Business Concept Development Stage seeks to establish a clear understanding of what is required to deliver the asset in its entirety and the corresponding client-side operational framework that is needed to deliver it. This will evolve into a client-side business delivery plan.

The Business Concept Development Stage provides an aligned approach to corporate project delivery

The corporation understands the corporate construction risk, has modelled the probability and likely scenarios at high level using appropriate tools and agrees the dynamic shift to ensure the appropriate risk management culture is in place and the corporation is aligned to success through a communication strategy.

Risk management is owned at board level and undergoes continuous review and development at operational level

This is a high level strategy signed off by the board permitting a more detailed understanding of what is required to provide **client delivery capability** at management level. Business development is an ongoing process, ideas will evolve and emerge as the developed operational plan proceeds, requiring board reapproval if there is a significant deviation from the business concept.

Note

1 Project management has a number of terms for clients. **Senior responsible user** (SRO) is used in government circles as a role with 'end to end' responsibility; owning the business case, ensuring the project meets its objectives and delivers benefits. **Project sponsors** similarly act as champions for the project, as escalation points for the project manager and a link between the business community and the project. The **intelligent client** is used in the Institution of Civil Engineers 'Intelligent Client Capability Framework'.

16b Corporate Client/Delivery Capability/Transformation Stage

16b.1 Purpose

<u>Corporate Client/Delivery Capability/Transformation Stage</u>: To develop the concept and argument for change into a more detailed framework of delivery with options and steps for the provision of a physical space, all within the context of the corporate appetite for risk.

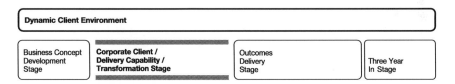

Figure 16b.1 Corporate Client/Delivery Capability/Transformation Stage

16b.2 Corporate Client/Delivery Capability/Transformation Stage

In the last stage we secured board level buy-in to our business case and *high level* client capability/de-risk strategy, a project board, a corporate governance structure and strategic framework. This section focuses on detailed planning at corporate management level required prior to design and construction activities. Plan for one year at this stage.

The **Corporate Client Delivery Capability/Transformation Stage** includes four 'supply side' outputs shown on our construction ERM (CERM) process model. Final sign-off to enable budget planning and delivery is at executive level.

(1) Client-side operational management plan systems and control framework (including client requirements/enabling soft and hard delivery statements).
(2) Micro cashflow planning and corporate risk within corporate planning.
(3) Procurement strategy (realignment analysis and supply chain management).
(4) Aftercare and operating policy.

16b.3 Job book: Corporate Client/Delivery Capability/Transformation Stage

16b.3.1 Operational planning outputs

Strategic decisions: sign off client-side operational management plan systems and control framework, micro cashflow planning

> **Corporate management activity:** *produce client-side operational management plan systems and control framework, micro cashflow planning and corporate risk within corporate planning, procurement strategy realignment analysis and supply chain management, aftercare strategy and operating policy*

Briefing: What are the packages of work we need to deliver the change, what skills are associated with the packages of work, how are we going to manage day to day finances so it does not impact upon the corporation?

Client-side operational management plan/systems and control framework

A client-side operational management plan is a more detailed brief of how the project is going to be delivered in its entirety. Construction design and delivery stage is one aspect.

Essentially there are three areas of client operational planning managed at operational management level:

- Configuration and delivery strategy
- Milestones and timeline planning
- Soft and hard client delivery objectives. These are:
  - Enabling works
  - Client requirements for the design and build
  - Operational delivery

The client may wish to employ a specialist process manager to manage these areas on their behalf.

<u>Configuration and delivery strategy</u>

The configuration identifies the individual packages of work (also called the workstreams), their order and the essential matrix through which the project will be delivered. Think of it as the 'jigsaw' of the project.

Packages normally describe:

- the output
- the method
- the skill set required.

The purpose of configuration is to describe the best logic to manage delivery and to optimise cashflow, to identify 'pinch points' and critical paths and to minimise risk. Configuration provides the optimum sequence of the project. It may have to be rebased several times throughout the project.

Configuration also provides the basis for the procurement strategy, that is to say whether the package is going to be provided internally or externally by a supply chain or by experienced internal resources. Make or buy decisions are therefore a subset of configuration:

Make or buy decisions

Having established a clear client brief and determined what is known and not known, this becomes a high level configuration problem.

There are two levels of configuration and they are both fundamental de-risking exercises to be revisited once the design team is appointed and the strategic brief agreed (Outcomes Stage).

Level 1 configuration: What is inside and outside the construction project and in which order is it done

The decision should be taken based on what is core to delivery of the business objective(s).

The three packages normally assumed to be outside the project are:

- <u>Works enabling the project</u> – that is to say getting the business and physical environment prepared for the construction project.
- <u>Client fit out</u> – that is to say; furniture, IT and graphics/ marketing.
- <u>Operational preparedness</u> (see 'Three Year In Stage') – that is to say enabling business operations to optimise the benefits in the business case and as delivered by the construction project.

Level 2 configuration: (inside the project) what does the client know and can control and what do they not know and what is the risk (specialist / architectural design)

Most design processes can be split into three risk areas, two client knowns and one client unknown. **The two client knowns are:**

- *No/minimal change* – what do I know, works well and does not have to change. In which case this process can be fundamentally designed by the client, should be agreed internally with minimal input from the design team.
   Design competence rests with the client and the client brief should make it clear what this process is and that no detrimental effects will be established by the process or the project.
- *Change driven by industry innovation* – what new processes/ equipment are being introduced by the client to drive change. Design competence lies outside the client and design team. The onus is on the client to make sure they and the design team are fully aware of the implications of the change and understand the impact on the design.

Both of these areas should be fully investigated before appointment of the design team and stated in the client brief.

The client *unknowns* are essentially the design driven by the layered process of the RIBA Outline Plan Stage of Work (see Appendix 2B) and will have an interface/change on the above.

Control of the design process (client red lines)

Having taken the decision in the configuration plan to externalise a package packages as part of the supply chain we need to think about what control mechanisms would apply to an external package. There are two different types of control:

- A control on design: client red lines, change procedure and corporate management.
- A control on design team performance: benchmarks and KPIs upon which the success of the supply chain will be measured.

The key focus on controlling design risk should be to establish what is *essential* and becomes the immovable client voice around which design moves and what is *desirable* and can be traded as part of design negotiation. These are the *construction critical success factors* and are agreed with the executive/board.

- The <u>essential and beneficial success factors</u> set the boundaries around which the client reviews any necessary design (and construction) change decisions to ensure business success.
- <u>Essential success factors:</u> the immovable client voice. These will be agreed with the executive/board. These are:

  - Accepted key business multiplier – that is to say the number of covers in a restaurant, the number of beds in a hotel, the range of treatments in a spa, that make the scheme profitable.
  - Viable operating cost or overhead.
  - Payback within investor norms.
  - Essential operating process – operating procedures must be enhanced and not put at risk.
  - Essential delivery – a date by which the project must be in operation.

- <u>Beneficial success factors</u> – achieved at the best possible cost and the <u>least organisational outcome risk</u>. When faced with the request for change the client should ask themselves:

  - Is the cost of design and construction within a reasonable norm expected for that type of project? <u>and</u>
  - what is likely cost to the organisation to achieve it. In monetary <u>and</u> business impact assessments.

Milestones and timeline planning

This will be an overarching indication of the broad timescale for the project, in descriptive terms without timescales, the latter being applied when further task and duration detail comes into the picture at a project planning level.

At its highest level a project is likely to consist of the following three stages:

- <u>The transitional period</u> – the implementation of soft (policy and managerial change) and hard (enabling) works to create the transformative environment to undertake the works.
- <u>The operational transformation period</u> – when the works for change come into place and the impact of change is felt on the corporation at an organisational and physical level.
- <u>The strategic transformation period</u> – change is delivered and permanent, the products have the greatest impact on the organisation and some moderation might be required.

A natural sequence of planning will emerge from the milestones (outline business case) and create a logical high-level sequence of project delivery, from this emerges the configuration and, subsequently, the project gantt chart creating the detailed task and timelines of the programme.

 Soft and hard client delivery objectives

Typically, in sequence, these are:

- Enabling works
- Client requirements for the design and construction element
- Client fit-out
- Aftercare and operating policy.

 Enabling works

These are works (demolition, site clearance and sterilisation, infrastructure, outside site boundary access and others) that will enable the project to be delivered by the client and costed outside the project. They can impact upon construction delivery and cause delay and incur penalties for the client.

Client requirements for the design and construction element including known constraints and statutory and planning permissions

This is a statement of how the client sees the facility operating after it is delivered and is an essential precursor to the Concept Design Stage (Stage 2) of the RIBA Plan of Work. It essentially covers:

- Why the client is doing the project.
- How the client sees the building operating in business practice.
- What is inside and outside the project.
- A schedule of the required floor area.
- The quality of construction and longevity required.
- The environmental and sustainability requirements.
- What the deliverables/interfaces will be.
- What the constraints to the project will be.
- What the drivers and dependencies will be.
- What is unknown and what the assumptions are.

 Client fit-out

That is to say, furniture, graphics/marketing, catering, telephones and IT. These will be directly managed by the client but will have interfaces. This is covered in the next chapter.

Aftercare and operating policy

This process includes:

- Handover of the facility and practical completion.
- Training and operation and maintenance manual/instruction and training.
- Aftercare regimes and user feedback.
- Operational staffing training and the development of operational policies and procedures to accompany change and to use the facility.
- The employing of any third parties/signing of leases/contracts to operate the new building, including any third party fit-outs under licence.

See: Aftercare and operating policy supplementary materials on our website.

Micro cashflow planning and cashflow controls

In the previous chapter we discussed establishing funding. Once the cashflow has been agreed at board level the client role is then to ensure good governance and accountability in the project. That is to say reducing corporate risk and exposure and establishing trust in the process by using the standard financial corporate mechanisms. This can be complex in a highly construction cost dynamic environment and needs finance staff that understand building contract payment terms.

Outer and inner funding control mechanisms

A typical approach would be to have two separate budget systems working alongside each. In effect two concentric circles. An outer ring attached to the corporate machine, with corporate dampeners containing the oscillations of an inner circle which fluctuates in response to the changes in the project.

The better the control within the inner project circle the less dampening required from the outside. However, they have different mechanisms and protocols. For the purposes of this book we will explain the main characteristics of the two systems.

Outer circle (corporate management)

This ensures the project is a manageable financial risk for the corporation.

The client responsibilities in a project are:

- To provide value for money in purchasing of consultancy and contracting.

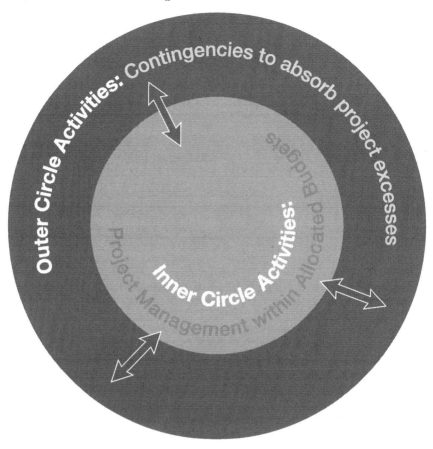

Figure 16b.2 Outer and inner circle financial controls

- To provide systems for overall budgetary control including checks and counter checks.
- To provide the standard finance functions:
  - Auditing – probity – proper management of accounts – commitment accounting and regular invoicing and ensuring contract payment terms are met.
  - Cashflow – impact of capital payments on corporate cashflow – can be significant and might affect project configuration.

These responsibilities can be discharged by behaviour throughout the project and by due diligence in funding.

The client must balance ambition with cost reality.

- Clarifying the fullest scope of works pre-tender.
- Distinguishing essentials from desirables.

- Managing change sharing the burden of difficult decisions.
- Due diligence in funding.

Essentially the outer circle actions are to provide funding for:

- Standard project exclusions. There should be a clear understanding from the earliest point in the project what is within/outside the project.
- Tax variances. Some elements of construction are VAT exempt. Most clients expect design teams/QSs to be experts in this field. They are not and the client should seek specialist advice from a VAT specialist.
- We would also allow two elements of client undeclared contingency (not revealed to the design team).

  - Exclusion interfaces. There are most likely to be elements that are unclear whether they should be provided within the scope of the works (normally wiring/data wiring that enable typical project exclusions like equipment, or existing client TV/fire alarm systems). About 5 per cent of the project budget.
  - Client variation contingency (This is additional budget and construction – undeclared to the design team – for unforeseen client variations – project users will want more as the project evolves. I would keep a 15 per cent contingency on the project budget until stage 4).

Inner circle (project)

In the inner circle the client transfers risk through appointment. The project budget will be agreed with the design team and sound budgetary management is within their professional competence. Overspends are an issue of non-competence.

Therefore, their role is:

- To ensure full and proper client requirements. To update the client requirements throughout the project.
- To appoint the appropriate expertise to discharge the client requirements within and outside the core design team. The above includes ensuring the appropriate level of design liability is assured.
- To enable the interface between individual client appointed experts.
- To create project heads and manage the project cashflow within allotted variances.

However, the client maintains a responsibility for proper **behaviour within the inner circle** and a monitoring role of the project budget controlling any volatility within the inner circle that might affect out circle stability.

A project budget cannot be developed without a scheme and without detail. Clients should expect the very earliest project budgets to be nothing more than the design team 'feeling out' the clients' comfort zone in terms of design and available budget.

At Stage 0–2 the client should allow a 20–25 per cent margin on whatever project budget is advised for fees and construction. QSs typically allow an additional 4 per cent for design variations right up to Stage 3.

Procurement strategy – *Market Testing, Realignment analysis and supply chain management*

The key to successful procurement is to be clear about what is wanted, for how long and why it cannot be done within existing resources. It is a make or buy decision.

The project will generally consist of three major packages, unless the contractor is a developer.

- <u>Client:</u> operational, enabling and client direct (specialist) works, furniture and fit-out/IT, furniture and loose fittings.
- <u>Design team:</u> Design up to Stage 3 or Stage 4 and contract administration.
- <u>Contractor:</u> Construction from Stage 5 *or construction and design from Stage 3 to Stage 7.*

In undertaking the packages, the client, design team and contractor all share risk and reward. The client ultimately takes the greatest risk and is responsible for the degree of supply chain integration and their performance. Different styles of construction contracting are covered earlier in this book.

This book does not go into detail on the procurement process. A public sector body must comply with European Purchasing Regulations if services or construction are over a certain value. This is a two stage process of public notice (PIN) followed by a selection and interview. Anybody in the EU can apply. The process of selection is strictly governed and procedures can be challenged – which may mean the competition has to be re-run.

The alternative is to use one of the many government frameworks which still use a two stage process but through a narrower pool.

Private sector – not bound by regulations but underlying principles must be fair and equitable. There is no requirement for a public notice and architects and contractors can be selected through recommendation.

Design competitions

Architects can also be selected by design competition. Managed by the RIBA, this not only complies with the EU directives but it also enables the client to select a number of architects to present completing designs for Stage 1+. All architects are paid and will require a developed user brief, but the client advances the design stage while complying with the regulations.

Contractor competitions

Contractor competitions will be based on the design and specification package prepared by the design team, either at Stage 3 or 4 depending on the level of quality control required. If it is a design and build contract, part of the selection criteria will be the confidence the client and design team have in the contractor's capability to achieve the design element, and how the contractor is going to achieve the performance criteria laid down by the design team.

<u>Sharing Risks</u> – the client's best strategy is to continually keep the risk owner under review – *risk is best located to the party best equipped to manage it* – and look at the benefit of risk transfer at any stage of the project, even if it means a contract change.

Realignment analysis

In reviewing returned contractor packages, taking into account feedback from the market testing process, having cognisance of the scopes and prices to ensure the project is being competed in the most pragmatic and cost effective manner. Necessary adjustments might have to be made to the client's operational plan.

Collaborative business relationships

Going forward the relationship should be governed by collaborative business relationship (based on BS11000). Again, a process manager could help develop this.

Common characteristics and behaviours are:

- Early visibility and commitment to pipeline of projects.
- Supplier relationships.
- Effective use of contracts.
- Encourages innovation.
- Set of assessment tools that allow sponsors, clients and supply chain to align behaviours and identify capability gaps.
- Pragmatic approaches to compliance.
- Next decision is the best procured solution – risks.

Aftercare and operating policy

This is a statement of how the building will be managed following handover and to a point at which the business case reaches maturity (usually three years), including asset ownership, as built information and BIM, management, operational service contracts and maintenance planning.

16b.4 Summary: Corporate Client/Delivery Capability/Transformation Stage

The purpose of the Corporate Client/Delivery Capability/Transformation Stage

The Corporate Client/Delivery Capability/Transformation Stage sets out the client-side operational management, cashflow and high level configuration and procurement plan for the project. It includes the aftercare and operational strategy.

The Corporate Client/Delivery Capability/Transformation Stage provides the framework for managing risk within asset delivery

Risk is managed by having a comprehensive plan for delivery within a budgetary envelope that has built on 'checks and balance' mechanisms with a strategy for managing the procurement and design process to ensure business success.

Design and construction risk management through a clear client brief and established business red lines

Design risk is managed through allocating known design risks to the best equipped party and having a clear and established strategy for conducting design stage negotiations.

16c Outcomes Delivery Stage

16c.1 Purpose

<u>Outcomes Delivery Stage:</u> To ensure the client vision and interests are maintained through design and construction through <u>pre-design and design client control measures (client knowns, essential and desirable change and benchmarking)</u> and to ensure the asset is optimised for handover to asset management stage.

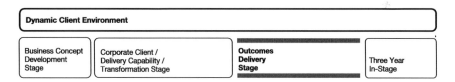

Figure 16c.1 Outcomes Delivery Stage

16c.2 Outcomes Delivery Stage

In the last stage we developed a pre-design strategy for the client and how corporations should plan for a project up to procurement of design.

The Outcomes Delivery Stage covers the whole of the RIBA Plan of Work from Stage 0 to Stage 7 as well as the construction stage. This would typically be up to two years. During this stage the role of the client can be summarised as:

- **Monitor client's interest** through pre-design, design, delivery and post-delivery stage.
- **Troubleshoot** – use best solution to resolve corporate and project roadblocks to keep it on track/budget.
- **'Flex' the output, benefits and overall project design** if necessary, to maintain success.

Manage client direct responsibilities

- Community infrastructure levy: CIL.
- Manage client direct contract packages.
- Manage land ownership/tenure issues.
- CDM health and safety.
- Safe working and access records.
- Manage planning risk: application for planning – pre-app Stage 3 –planning consultant.
- Professional contracts (see post note about design liability).
- Third parties – collateral warranties, step in rights.
- Insurance – notify insurance.

16c.3 Job book: Outcomes Delivery Stage

16c.3.1 Asset delivery outputs

Strategic decisions: sign off Outcomes Delivery Stage, including benchmarks, updated benefits realisation, asset receipt strategy

Corporate management activity: *produce Outcomes Delivery Stage including undertake client's legal duties and monitor interests, troubleshoot and flex as below, produce reviews of performance against benchmarks and updates on budget costs impacts on key business impacts and achievement of benefits plan*

Briefing: How are we staying within the benefits plan during design and construction, what are the major variations, how do they affect the business outcome, how are the design team performing?

Pre-design, design and start on site

This section is split into:

- The client and the design and construction process in overview.
- Preparing for design, benchmarking and design liability.
- Appointment of contractors and start on site.

RIBA stage	RIBA tasks (related to clients only)	Client activities (monitor, troubleshoot and flex)	Client control tools (monitor, troubleshoot and flex)
Appoint process manager[1]			
Pre-design		Client requirements and decision on: • Level 1 configuration: What is inside design scope/ outside design project.	Benchmarks: • Design team work stage completion against programme.

RIBA stage	RIBA tasks (related to clients only)	Client activities (monitor, troubleshoot and flex)	Client control tools (monitor, troubleshoot and flex)
		• Level 2 configuration: What is inside the project and to be designed by the client and programmed by the design team. • Level 3 configuration: Client key control measures of design: essential measures of value and key red lines.	• Building design cost/m_2, energy usage/m_2. • Amount of work actually measured at tender. • Planning officer recommendation. • Accuracy of QS estimates against tender returns.
Appoint project manager. Appoint quantity surveyor, appoint architect[2]			
Stage 0: Strategic Definition	Understand business case, client strategic brief. Understand planning and programme .	• Agree Level 1 and 2 configurations. Introduction of measures of value and key red lines. • Agree soft landings. • Agree process for appointing other design disciplines as design evolves.	• Agree and test benchmarks. • Appoint project manager and quantity surveyor as separate appointment from design team.
Stage 0–1 Undertaken at quantum merit rate.[3]			
Stage 1: Preparation and Brief	Agree project objectives, project execution plan, initial project brief scope of service. Design responsibilities and BIM.[4]	• Site information, risk and early feasibility studies test and refine clients' measures of value and key red lines. • Client goes back to board/refines business case.	• Client establishes gateways. • Training of client for BIM.[4] Client writes employers' information requirements as PAS 1192-2.
		• Initial project brief. • Agree client responsibilities under design responsibility matrix. • Prepare funding strategies.	• Client agrees project execution plan, establishes project board. • Client manages common data environment.

Outcomes Delivery Stage 263

RIBA stage	RIBA tasks (related to clients only)	Client activities (monitor, troubleshoot and flex)	Client control tools (monitor, troubleshoot and flex)
Finalise architectural and design team contracts[5] Ongoing fee established as percentage of net construction value[6]			
Stage 2: Concept design	Outline concept, project brief, procurement and construction strategy.	• Receive/sign off Stage 2 report with preliminary design(s) and budget costs. • Reconciliation of key red lines with design including structural mechanical and electrical designs. • Reconciliation of project design with client designers, in preparation of client interfaces. • Agree the style of construction contract and how construction will be procured.[7] • Finalise funding strategies.	• Is the design team making good progress at this work stage completion against programme. • Consider early building design cost/m^2, energy usage/m^2.
Stage 3: Developed Design	Developed design, planning application change control.	• Receive Stage 3 report[8] with preliminary design(s) and budget costs. • Seek approval of board to Stage 3 design against established client key control measures and red lines. Update on benefits and business case progress.	• Last opportunity pre-tender review of design team work stage completion against programme. • Last opportunity pre-tender review mid stage building design cost/m^2, energy usage/m^2.
		• Sign off design against agreed Key Performance Indicators (KPIs). • Consideration of construction contract procurement.	

RIBA stage	RIBA tasks (related to clients only)	Client activities (monitor, troubleshoot and flex)	Client control tools (monitor, troubleshoot and flex)
Sign off design team fees (approx. 30 per cent)			
Stage 4: Technical and Design	Technical design, sustainability, maintenance and operational and handover strategies, submit building regulations.	• Receive and review pre-tender documents. • Receive budget cost report and agree/direct value for money savings.[9] • Authorise tender release. • Prepare and align organisation locally and at each level for project. • Bring funding strategies into operation.	• Review benchmarks: • Design team work stage completion against programme. • Building design cost/m^2, energy usage/m^2. • Amount of work actually measured at tender. • Planning officer recommendation. • Establish client time and budget contingency.[9]
Sign off design team fees (approx. 60 per cent)			
Stage 5: Construction	Tender stage, start of construction, contract progress and prepare for handover. Contractor insures building site.	• Receive tender report from quantity surveyor. • Review against budget cost and funding strategies. • Review leading construction company accounts. • Monitor project progress against stage completion.	Benchmarks: • Accuracy of QS estimates against tender returns. • Manage variations in construction that will have time and budget implication.
		• Manage change/variations at the board. • Manage client's direct works within contract programme. • Sign construction contracts and construction design liabilities.	
Stage 6: Handover and Close Out	Covered in next section		

RIBA stage	RIBA tasks (related to clients only)	Client activities (*monitor, troubleshoot and flex*)	Client control tools (*monitor, troubleshoot and flex*)
Sign off design team fees (approx. 97.5 per cent)			
Stage 7: In use	Covered in next section		

Notes
1. A process manager is a more neutral appointment that evolves the business case without being constrained by a construction background.
2. A project manager is a project construction specialist. The architect as lead designer and design team leader is the first appointment. Other consultants are appointed as the design evolves.
3. Quantum merit (hourly rate) until initial project brief is established.
4. Building Information Management level 2 (BIM) should be managed by the client as a way of identifying and managing risk between stages.
5. Architectural and design team contracts should be fully concluded when the full design responsibilities, design responsibility matrix and design liabilities are fully understood under the contract. See design liability.
6. Once a net construction cost can be established, design team fees are agreed as a percentage of the net construction value.
7. Construction contracts can be design procure and construct or a design and build contract in which the contractor undertakes part of the design with a novated team undertaking a part negotiated process.
8. Often considered the last stage the client can make comments, potentially client latent operational/design issues emerge for first time. This can be the cause of variations and in some cases the client fundamentally revisiting design.
9. Clients should plan to have a confidential additional time and budget allowance over and above that stated within the design report.

16c.4 Preparing for design stage, benchmarking and design liability

Project manager

The first construction appointment is likely to be a project manager either from the Royal Institution of Chartered Surveyors (RICS) or Association for Project Management (APM) – both of which are recognised professional bodies of knowledge whose members carry the appropriate professional indemnity. They manage the packages inside and outside the project on behalf of the client, the design team appointments and the project budget.

<u>*You are the only one that understands the business.*</u>

Project managers are often considered as a quasi-client. This is a misunderstanding of two fundamental precepts:

- The project manager's ultimate goal is to deliver a construction project and their specialism lies in packaging up and managing projects for delivery. They are not business specialists. That is why we recommend a process manager at the Business Concept Development Stage.

- The project manager cannot evolve your goals, objectives and benefits as business opportunities arise during the course of design and construction, which will happen, and only an intelligent client is equipped to do so.

However, a project manager can assist the client in the earliest (pre-RIBA) stages of project development and can overlap with a process manager.

Architecture/design team

The architect has two functions in design:

- Lead designer – that is to say being the primary concept developer, upon which other design (structural/mechanical and electrical and specialist design) is subservient.
- Design team leader – essentially a management function that ensures all subservient designs are delivered in a timely way, are integrated with the main design and other designs and the interfaces are resolved. In effect ensuring a complete design solution is offered to the client. We will return to the issue of design liability later.

The client should ensure that both roles are carried out effectively.

Establish benchmarks/client redlines: how to work with a design team; monitoring your (client's) interest through design and construction

The design and construction package will be one of the multiple workstreams managed by the client to complete the business case but it is the highest risk and will have the greatest impact on the corporation during its development.

As the design develops, the client requirements drafted in the Corporate Client Delivery Stage will be merged with site constraints and statutory design criteria to form the strategic brief for the design. The result is inevitably a compromise with some compromises being made on the client's business objectives as well as other opportunistic benefits emerging as the final design becomes clearer.

Client dissatisfaction in construction is mostly not caused by change itself but the manner in which it is achieved. Disempowerment and the inability to control or moderate the output is a common complaint. From the design team perspective, clients are often cited as changing their mind, deliberating and causing time and cost impact.

Our RM process addresses these concerns by getting the client at the pre-design stage to determine:

- What they want to control directly in design and construction.
- For what they can't control – they determine clear values 'red lines'.

Those elements outside the project will be the least risk to the client (i.e. enabling and furniture). There will be programme and interface issues with the project design and construction team (these are generally managed by the project manager). Similarly, some client design elements will have to be installed as part of the project. Most building contracts allow for this as clients' 'artists and tradesmen' who are included within the contractor's programme but otherwise responsibility for their output rests with the client.

Ensuring business objectives are achieved at design and construction stage

The client requirements and business objectives go through most change through the development process that forms Stage 0–3 where the client's precious 'red lines' are merged with a number of other considerations including site constraints, planning consent and building control to form the strategic brief.

Managing design risks during design development:

- Design **Risk One**: Feasibility Stage (Stage 0–1) is often achieved at considerable compromise and at great cost to the unprepared client. At great cost because this stage is billed at an hourly rate (quantum merit or QM) and can be open ended.
- A client that is unclear about what can be compromised in design and that does not have the ability to justify to the board that design has optimised the critical success factors/essential measures of value laid out in the client requirements runs the risk of an inconclusive and expensive design Stage 0–1.
- Design **Risk Two**: At Concept Design (Stage 2) the fee becomes a fixed percentage of the construction cost; however, client concept ideas are still emerging at this, and later stages, therefore architectural contracts usually allow the right to charge a QM rate even if a percentage fee has been agreed.

The cost of changes at design stage should considered against the benefits plan and positive changes justified to the board

These risks and the client's interest can be substantially mitigated and managed at design stage by using the techniques previously set out in this book, and, if required, by a trained process manager. As discussed earlier clients should go into the design stage having established the client knowns and a strategy for controlling the unknown (or design risk) through critical success factors/essential measures of value. They are summarised below and discussed in full in our earlier chapter. Both areas should be fully investigated before appointment of the design team and stated in the client brief.

👍 *Establishing the client 'knowns' (level 1 and 2 configurations from previous chapter)*

- No/minimal change – what do I know, works well and does not have to change.
- Change driven by industry innovation – what new processes/equipment are being introduced by the client to drive change.

👍 *Client design control measure 2: controlling design risk by establishing essential (inviable) and <u>desirable</u> (what can be traded as part of design negotiation – level 2 from previous chapter)*

In the last chapter we outlined the essential and beneficial change controls.

- <u>Essential measures of value:</u> the immovable client voice around which design moves, that is to say essential products of income and essential delivery dates.
- <u>Beneficial change</u> achieved at the best possible cost **and** the least organisational outcome risk. When faced with the request for change the client should ask themselves:

  - was cost of design and construction <u>within a reasonable</u> norm expected for that type of project **and**
  - what's the likely <u>cost to the organisation to achieve it</u> in monetary and business impact assessments.

- **Design Risk Three** is that planning permission might be refused at Design Stage 3. Having established the strategic brief, this is essentially a design stage risk. This is mitigated by having pre-application(s). This is a dialogue with the planners before the planning application is submitted about the intended design. It does not guarantee planning permission but reduces risk. A further measure is a planning agreement by which the client and Local Planning Authority (LPA) agree (for a consideration) a programme and dedicated resource to manage a timely planning process. Again, it does not guarantee a successful planning application.

👍 As part of the client requirements the client should have developed some benchmarking system to measure design team performance and manage the design stage. Typical benchmarks and key performance indicators are:

- Design team work stage completed against programme (*flush out performance issues with lead designer and design team leading*).
- Building design cost/m$_2$, energy usage/m$_2$ (*a measure of effective building design*).
- Amount of work actually measured at tender (*a measure of tender package quality*).

- Planning Officer Recommendation (note: it is not possible for an architect/planner to guarantee Planning Approval, only get a recommendation of approval from a planning officer).
- Accuracy of QS estimates against tender returns.

Architectural and construction contracts and design liability:

Like any other industry construction is a network of design and manufacturing collaborations, some of which are inside the project and some of which are outside, that come together to form a project.

Fundamentally the major difference, particularly in the UK, is that in all contracts design and production is carried out by separate parties by common agreement.

Design is the greatest risk in construction and can be undertaken by the client's direct subcontractors, the design team and the contractor (in a design and build contract).

The design liability is defined by the standard professional agreements (contracts) of the professional institution (RIBA, ACE, RICS, APM) and the individuals designer's/contractor's insurance.

If the contractor takes responsibility for any part of the design, they must have the appropriate design indemnity and liabilities included as part of the building contract.

Limitations on design liability

Design liability is normally limited to 12 years after the date of the discovery of the defect.

The RIBA standard contract has a net contribution clause – in essence this limits the architect's liability to their part of the design, even when the architect is paying all other designers. This can come as a shock to clients who expect, in the event of a negligence claim, the architect to accept liability for the whole design as a 'one stop' shop for design liability and then claim costs from other designers retrospectively.

This is not the case and at best the architect will assist the client in establishing which of the design team is responsible. This is particularly problematical as design by its nature is normally integrated and often design liability takes many years to establish.

Engineering contracts (such as ACE) generally restrict professional negligence claims to a multiple of the fee, therefore if an electrical engineer is proven to have made an error on a £40k fee, typically their design liability would be restricted to ten times this fee, a maximum of £400k whatever the cost of the defective work to repair.

An indicator of design liability is the design responsibility matrix, usually a document the design team publish amongst themselves at Stage 3/4 to determine who is responsible for the interfaces. This should be part of the architectural contract signed by the client.

Beyond this the client may ask for a bespoke contract or an addendum to the RIBA Appointment that says the architect is responsible in the first instance. This would require the client to cover higher insurers' premiums and possibly a reduction in the term of liability.

It is for the client to judge the relative cost and most effective way of managing design liability.

How Building Information Modelling (BIM) may help with tracking design liability

A further aid to tracking design liability is Building Information Modelling (BIM). Usually considered as a design tool of not great value to non-construction clients, it has emerged in recent studies as a possible record of inter-stage risk: that is, risk as a constantly evolving entity, with emerging and non-linear aspects that move between different project stages potentially to the benefit or disbenefit of the successive owner, can be captured and tracked through BIM.

The client is required to express what level of BIM is required on a project, the usual request is for BIM level 2, at build not create level. This is the most fundamental level of design requirement.

16c.5 Appointment of contractors and start on site

Post-Design Stage 4 a tender pack will sent to competing contractors. The quantity surveyor will receive contractor's prices and make a recommendation. The client issues on tender return (market testing) will be:

- The range of price returns and whether there is a favoured contractor and a first reserve within affordability.
- The financial standing of the proposed contractor and their financial rating/solvency.
- Gearing up the corporation for the works and pre-contract planning – see enabling works.

The building contract will state a start on site date and practical completion date. The contractor will require a 'lead in' period to mobilise and eventually the site gets handed to the contractor to undertake the works specified in the tender, the second is the date by which all of the works are deemed complete and the site gets handed back to the contractor. Both dates have legal significance supported by case law. During this period (RIBA Outline Plan Stage 5) the construction site is managed and insured by the contractor.

Contractor design portion (CDP) appointment

Where a contractor is appointed to design, work may not start on site immediately. The contractor's specialist subcontractor will undertake a design stage and present 'shop drawings' for the design team to approve. The contractor will have the same design liability as the design team and it will be set out in the building contract.

Client culture in design meetings and site meetings

The client has a unique role in construction projects and how the client as project leader behaves will ultimately determine the character of the project.

The client as developer gives the project a business concept, and develops the financial shape. The client is also an end user with life cycle and quality considerations. They will also fulfil the role of public relations advisor, primary budget manager and risk reducer including potentially managing bankruptcy/statutory and always having an exit strategy.

However, during design team meetings, the client is team builder, key health and safety executive, benchmark and quality auditor and tone setter, looking for positive relationships, continual improvement, best practice in ecology, respect.

16c.6 Summary: Outcomes Delivery Stage

The purpose of the Outcomes Delivery Stage

To ensure the client's core vision and interests are maintained while troubleshooting and being flexible about the benefits and overall design. Design is an evolutionary process and ways of delivering the client's stated values may change.

For the best outcome manage your design and construction team with clear guides to success, ensuring your interests are maintained, as a high performing client

The best outcomes are achieved by clients who are clear and consistent about their objectives (*client red lines*), have a clear process of selecting talent and how to optimise it during delivery and are clear about how they can add value to the process.

Success is not just measured on the outcome but the way by which it was achieved.

The design process will determine asset performance and value

Ultimately the design and construction process is an important stage of the asset lifecycle and variations in design should be measured against the overall cost/benefit of the business case.

16d Three Year In Stage

16d.1 Purpose

Three Year In Stage: To ensure the client vision and interests are considered at a stable operating year of the asset and whether any further action is required to improve asset performance.

16d.2 Three Year In Stage

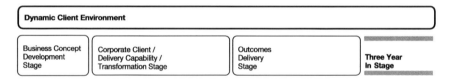

Figure 16d.1 Three Year In Stage

In previous chapters we have looked at how the client's interest is monitored throughout design, how issues of conflict between the design and business objectives are resolved and how the client develops the benefits plan in accordance with design development.

The Three Year In Stage covers RIBA Plan of Work Stage 6 (Handover and Close Out) and Stage 7 (In Use). However, we take it further:

- These stages are critical for asset management. Stage 7 is effectively the post-completion stage during which outstanding documentation is handed over and a post-occupancy review of construction takes place.
- The asset is there to support business objectives. **Therefore, in this book we also focus away from design to how asset performance at handover supports the business case, which gave rise to its development.** This is the 'acid test' of a successful project.

The Three Year In Stage starts at construction handover. It is equivalent to RIBA Stage 6 and Stage 7 but goes beyond technical delivery and merges business objectives with asset performance.

👍 **See supplementary materials: asset management and maintenance planning for clients on our website.**

16d.3 Job book: Three Year In Stage

16d.3.1 Aftercare and management outputs

Strategic decisions: sign off handover and acceptance of completed asset, review three year results

Corporate management activity: *produce asset handover management plan and corresponding client business development activities at handover. At Three Year In Stage undertake a three year in contract and three year in business review.*

Briefing: Have we got a seamless approach to asset and operational business management from Day 1? From our normalised year of operation 'three years' in, is the asset performing to expectations, have business benefits been realised or what needs to be done to improve the situation?

There are two parts to the Three Year In Stage:

- Part One: asset handover management plan.
- Part Two: post-operation asset review.

16d.3.2 Part one: asset handover management plan

Practical completion (PC)

The building contract will state a practical completion date. This is the date by which all of the works are deemed complete and the site gets handed back to the contractor. Both dates have legal significance supported by case law.

During the contract period the contractor is responsible for insuring the site and at practical completion the completed works are handed back to the client who takes over the legal and regulatory duties related to its safe management and operation. Practical completion is signified by the issue of a practical completion certificate.

Once the contract administrator issues the PC to the contractor any risk associated with the building being complete passes to the client.

Three week lead-in to practical completion

The handover of a building that has potentially (on average) taken up to three years to design and construct is in itself a process that needs management, as each element of construction and building system has to be transferred from the constructor to the operator. This is simultaneously a transfer of technical operating knowledge, and trust that it has been installed and commissioned correctly and this will take time.

A client training and operating plan as well as operating policies for the new building will be put in place by the client alongside the contractor's 'path' to practical completion.

A good project manager or architect should have at least a three week lead-in and should publish a closing programme listing every item and system with a process of handover. At PC the contractor's liabilities end. All but a sum equivalent to 2.5 per cent of the contract value is owed to the contractor.

Aftercare agreements

An aftercare agreement, whereby the contractor immediately addresses issues after practical completion, is essential to client/contractor trust, but not a contract requirement.

Patent and latent defects

A ***patent defect*** is a defect that becomes apparent during the course of the construction period and a ***latent defect*** becomes apparent after the rectification period/defects liability period (DLP) has passed. Patent and latent defects are discussed below and on our website.

Patent defects are addressed by contract procedure up to practical completion. After PC, and during the rectification period/DLP (see below) if a patent defect arises the project manager must decide whether it is a patent defect or maintenance (outside the scope of contract). If it is the latter the project manager will ask the client how they want to deal with it at an additional cost.

Rectification period/Defects liability period (DLP)

The rectification period/defects liability period is a period of six months to one year, established by the building contract, during which the contractor has to make good any 'shrinkages or other such defects' arising from the works. If the contractor fails to do this, then the retention (2.5 per cent) is used to complete the works. The end of this period is signified by a making good certificate. This is in effect the end of the contractor's construction obligation to the client.

Fit-out and 'turnkey' projects

A building contract is for a building structure and external envelope with all internal finishes and joinery and its associated mechanical and electrical systems.

The starting presumption in most projects at practical completion is that the building 'fit-out', that is to say furniture, IT installation and hardware (including servers), even signage, is excluded from the contract to be done by the client.

Some elements are debatable: for example, usually IT data wiring up to boxes – cover plates and 'patching' (the final connection of the data outlet to the server) are a client installed item and security and communications can be installed by contractor or client.

If the client wants a building that is fully complete, ready to occupy on 'Day One' (even down to the linen on bedsheets in a hotel) then the client must state it is a 'turnkey' project and the contractor undertakes 'fit-out' as part of the contract.

The scope is usually clarified in the client requirements and Stage 0/1.

Move management

The movement of occupants occurs after building handover is normally planned during the three week lead-in to practical completion. The company engages a move manager, packing crates are delivered and personnel's personal effects boxed for transportation and labelled with their destination. Tips are to communicate endlessly, agree a single point of occupant dialogue and a lost property store to place goods when labels become detached.

Client contract activities	Client business development activities
Handover	Receive handover of contract start. Fit-out contracts and soft facilities.
1. Agree soft landings plan with design team.	1 Fit-out (can be done by contractor). • Loose furniture • IT-switch plates, patching[2] • Computer hardware • Computer racking • Signage • Final clean
2. Prepare to receive building/notify insurers.	
3. Manage interface between client design systems and new build, i.e. IT and any installed security/cash, other systems.	
4. Undertake site visit with design team, attend design and site meetings, seek clarification when systems will be ready for commissioning.	
5. Clarify with both design and construction parties what success looks like, i.e. under what terms building handover will be accepted.	
6. Appoint external commissioning.	
7. Review architect's snagging list. Make client's contribution (non-contract items resolved as additional client instructions or after construction works).	
8. Agree aftercare documentation: • Operational manual review • Testing and commissioning • Statutory inspections • Health and safety file • Warranties • Planning approvals • As built drawings reviews • Energy use data • Energy performance certificate	2 Soft facilities: • Extend existing maintenance system contracts. • Catering installations. • Systems training. • Revise scope of maintenance contracts, i.e. cleaning and security. • Appoint additional staffing and develop staff training protocol. • Agree operational and maintenance protocols for contractor installed items. • Client takes back insuring liability from contractor. Advise insurers. • Client pays rates and service connection billing. • Ensure maintenance contracts for new systems.
9. Builders clean	3. Move management

16d.3.3 Part two: post-operational review

Latent defects

A **latent defect** becomes apparent after the defects liability period (DLP) has passed. The end of the DLP and the issue of the certificate for making good signifies the end of the contract. If a defect is not raised during the DLP there is no right under the building contract for it to be repaired.

Other potential remedies are:

- Breach of contract. There has to be a direct contract between the two parties and there is a limitation of 6 years from the end of the contract or 12 years under seal.
- Negligence. There is a limitation of six years from the date the action occurred.
- Claims can also be made under the Defective Premises Act if it is a premise.

Clients can also make arrangements before the contract starts through latent defects insurance, although its value should be considered case by case.

Post-evaluation

There are two types of evaluation:

- Three year in contract review.
- Three year in business review.

Three years is the normal assumption of a stabilised year of trading, when the asset will be operating at full capacity.

<u>Three year in contract review</u>

This is a review of whether the project was a success, very much centred on the way it was achieved rather than whether it was delivered to cost and budget. The purpose being to improve the process of delivery. It would also cover issues such as thermal comfort, acoustics, comfort, solar gain and other client occupational issues within the new asset and whether this needs to be improved.

<u>Three year in business review</u>

This is a review of the project's non-financial and financial benefits and whether they met the original objectives or were improved or compromised by the process of asset design and construction. This will look at cost of operations in use.

The decision may be taken to undertake improvements if necessary.

Client contract activities: Three year in contract review	Client business development activities: Three year in business review
Defect liability period 1. Raise defect issues with architect. 2. Review operating manuals alongside actual practice make amendments before making goods ends. 3. Ensure operating manuals reviewed by new staff. 4. Ensure all 'As built drawings' received. 5. Adjust handover settings for air conditioning/heating for summer/winter operation. 6. Adjust other handover settings for 'in use' operation. 7. Monitor building use and energy performance. Three year in contract review 1. Description of need delivered. 2. Performance of cost quality and timescale targets. 3. Client satisfaction with project and facility. 4. User satisfaction with the facility. 5. Performance and communication between project participants. 6. Overview and recommendations. 7. Technical appendices: user survey data /monitoring data. Asset maintenance plan 1. Service regime and costs (CIBSE). 2. Replacement costs and year.	1. <u>Actual surpluses/losses from asset in use years 1–3</u> • Actual Key business multipliers provided (covers in a restaurant, beds in a hotel, treatments in a spa, in operation. • Actual Operating Cost or overhead • Actual Operating Costs-. • Actual Payback <u>Against</u> 2. <u>Planned/Viable surpluses/losses Asset in Use Years 1–3</u> • Planned Key business multipliers. • Planned Operating Cost or • Planned Operating Costs/overhead. • planned operating costs. • planned payback. 3. Performance of asset against benefit plan • Reasons for under/overperformance. • Action plan. • Lessons learned.

Note: CIBSE = Chartered Institution of Building Services Engineers.

16d.4 Summary: Three Year In Stage

The purpose of the Three Year In Stage

To ensure the client's core vision and interests are delivered at hand over and through the defects liability period and to determine whether any further action is required post-handover to improve asset performance at its stabilised operating year (Year 3).

At building handover the client ensures that sufficient time and resource is allowed to receive the building as a full operating asset

A three week lead-in should be allowed to ensure all building systems have been full tested, staff are properly trained and all associated projects are complete to operate the building at optimum capability.

After three years a contract and business review should be undertaken to reflect whether any further work is required to maintain the asset at its intended performance capability

For the client, the building starts and finishes as a business venture. Design and construction is the first stage of an ongoing process of assessing how asset delivery and performance has and will continue to support improved business delivery. A review and further changes may be required in its first year of stabilised operation.

17 Key points: client risk management and the risk management model

(1) Clients have a natural stability developed from structures and protocols that suit their core activity. A construction project is a change initiated for a business purpose. Construction projects draw us away from our naturalised ability and cause a transfer gap in knowledge.

(2) However, construction is a risk reward venture undertaken by a client, a designer and a contractor. It is a fact that the party that takes the greatest risk, the client, has the least knowledge of the culture and processes of the construction industry.

(3) As a '*quasi-developer*' the client takes on the role of not only developing the business idea but also choosing the right resources and creating the right environment to successfully and seamlessly deliver it.

(4) When undertaking a construction project, the client will simultaneously go through a steep learning curve of understanding a well established culture in the construction industry, which will be different from what they know, as well as change in their own organisation that is initiated by the project.

(5) The design process itself can cause them to reflect upon their own initial business proposition, or maybe an underlying idea provides a beneficial change of direction at a critical stage of the project. Handled positively this is a positive process, mishandled it can derail the project and cause distrust.

(6) Such is the scale and impact of a construction project upon an organisation that it can be akin to a planned major impact. Delivered successfully, it can enhance the reputation of an organisation. Delivered poorly, it has an impact upon the competence of the organisation beyond the scope of the project itself, creating a **corporate construction risk**. Success in construction is not only measured in the outcome but the manner in which it was delivered.

(7) The greatest risk is *poor preparation*. 'Zero preparation', in the words of Bent Flyvbjerg, 'is as bad as it gets' (Harford, 2018). The client pre-genesis phase starts one to two years before an architect is engaged and is critical to reducing project risk; how a client configures the project, establishes the 'red lines' critical to project success and lays the ground for competence within their own organisation, will have consequences at each stage of project delivery.

(8) This is where the client is on their own, without the collaboration of designer and contractor and with the incumbent responsibility to get it right for when the formal design process comes along.

(9) Also essential to de-risking is an experienced delivery team. Clients will be the least experienced with *no body of knowledge* to call upon. They must rely on the experience of others to form a clear idea of how others expect them to behave and contribute to the process and a rationale that is likely to get them to where they want to be with least exposure to risk. One thing is for sure is that they must transform dynamically to become a 'professional developer'.

(10) What is a client's role and where does a client start to discharge their responsibilities? Entering into the world of construction design and production the client will find an industry with an established protocol and language. Nevertheless, as project leader the client must maintain the *open mind and flexibility* that is characteristic of the business world. Four behaviours are required uniquely of the client throughout the project:

- Clarity in purpose.
- Realism and flexibility in operational planning and change.
- Consistent client advocacy in design and production.
- Balanced management of change and corporate risk throughout asset delivery and stabilisation.

(11) There is not one single established approach for clients in construction, however many clients use remarkably similar approaches. This can be formulated into a single logical sequence which is characterised by our construction enterprise risk management (CERM) model.

(12) Our model starts with the **Business Concept Development Stage**, which embeds the idea of a project – an achievable vision – that needs a construction element, within corporation planning and thinking. At this stage the project is given a corporate shape by setting a framework and guiding principles that need to be delivered without the benefit of a design.

The outcomes are the expected *business benefits* delivered by a business concept (*a value proposition*) within the boundaries of an effective business delivery environment (*critical success factors*), set within an expected sequence (*a configuration*) and with predetermined delivery stages (*objectives and milestones*). Funding and cashflow will determine the date of a safe start, the reasonable duration given technical possibility and the date of handover within the corporate body's risk profile. A project board is established to provide a direct single point link between the project and the executive and governors of the corporation. At the very earliest stage a communique is used to advise the wider corporate body that a project is commencing, but also to align them for change, a thread of communication that is continuous throughout the project and essential to motivating the multiple strands of change that have to take place within the body corporate.

(13) The second stage, the **Corporate Client/Delivery Capability/Transformation Stage,** develops the principles agreed at the Business Concept Development Stage into a recognisable '*solution driven*' client-side plan for delivery. Work packages are devised – containable 'chunks of work' with a deliverable scope – so that problems with one package do not become a problem with project delivery. This is supported by a rational sequence of delivery, a broad programme of work and an outline delivery organisation with the capability and skills required to deliver the project. This includes the client body themselves and any gaps in their skill set at corporate management and strategic level.

The client will be required to develop their *client requirements* to give to their supply chain, establish a fair and rational process of procuring the right support skills and develop measures that the client will use to develop a successful aligned and integrated supply chain amongst the various contracted parties. This stage also sets out the operational management of the facility and proposed aftercare arrangements following handover.

Alongside the operational plan is a micro cashflow plan, with inner circle (project) and outer circle (beyond project) corporation controls, the latter acting as a dampening mechanism, absorbing any shocks that exceed the normal contingencies established by the project team.

(14) The next stage is the **Outcomes Delivery Stage.** At this stage via a negotiated design process, the client's business vision undergoes transformation to become the design *strategic brief*. This first formal stage of the RIBA Outline Plan of Work is the start of an established process that will create a built form that the client is obliged to accept as the construction element of their business vision. There will be compromises, however, with realistic 'red lines' and established *critical success factors/essential measures of value*, the final design should be able to be related to the *Benefits Plan* and to the original vision, at tender stage and at completion. If compromises are made then the board, who have agreed the business case and the red lines, has to be consulted. Otherwise, working within those boundaries, there is no reason why the *intelligent client- developer* should not be given autonomy to act in the corporation's best interest. Too many variations may be the result of an unrealistic vision or an unchallenged design.

Benchmarks are an essential measure of the performance of the design team but they also provide an 'early warning system' that client intervention within the design team process is required. However, clients should be cautious and realistic about design liability limitations embedded within standard industry professional contracts.

(15) Finally, the **Three Year In Stage** starts with an optimised handover, that aligns the delivery of the building asset with operational readiness and ensures the least risk of *patent and latent defects* arising.

The building at handover is not finished. A defects liability period ensures that technical faults are addressed within the contract, however this will not ensure the success of the business model or that the asset and operation is working together at its optimum capability.

There is no contractual remedy or prescribed review for this stage. Stage 7 of the RIBA Plan of Work focuses on the stage just after handover of the asset. We chose **three years** as our optimum point because it is the point at which most businesses take as their ***first stabilised year of activity,*** the point at which true surplus and subsidy are calculated, and the point at which the success of a business venture is measured.

Consequently, we suggest a *three year in contract review* and corresponding *three year in business review* to ensure the original aspirations of the business plan are revisited and, if necessary, changes are made to the existing asset.

(16) Fundamentally a building project starts and ends as a business venture. Great architecture helps market and draw attention to the proposition but in itself cannot make business success and failure. Some businesses look to the design team to develop the business idea.

(17) This is a fundamental mistake as it suggests a vague business proposition with a lack of clarity about how the business must operate effectively. Design teams work best when they are building on a clearly thought through idea and they are challenged to maximise its impact through design. Great solutions often come through design competitions, supported by comprehensive client recommendations.

(18) Skill, experience and time for planning is essential to success. The developer client is a unique skill currently unsupported by the *body of knowledge* available to project managers, designers and contractors. Part of the purpose of this book is to start the debate on how this can be developed. We have made a small contribution through our website.

(19) Allowing time for planning should not be underestimated. Our collective experience is that most projects are in the client planning stages (*Business Concept Development Stage and Corporate Client/Delivery Capability/Transformation Stage*) at least two years before any formal architectural advice is obtained. A significant part of our system takes place before the first stage (Stage 0) of the RIBA Outline Plan of Work.

(20) During that time the client is very much on their own. We hope that during this period our book and website act as a useful compass on how to move forward and manage risk.

Reference

Harford, T (2018). What the Sydney Opera House can teach us about Brexit: The rules for delivering a successful megaproject are devastatingly simple. *Financial Times*, 26 October.

Appendix 2A: Client-side CERM model

CLIENT SIDE PROCESSES

	Business Concept Development Stage	Corporate Client / Delivery Capability / Transformation Stage	Outcomes Delivery Stage	Three Year In Stage
Strategic Direction (Outputs)	Define Vision Value Proposition; Create Corporate Governance; Outline Business Case, Objectives, Benefits Management, Business Redlines	Create Optimum Delivery structure; Create Client Infrastructure; Reduce Corporate Construction Risk, put Dynamic Capability in place and controls system	Design / Negotiate benchmarks around emerging design; Construct Asset / Refine Asset; Business Case Refinement / De-Risk and Benefits Realisation; Define and De-Risk Asset Receipt	Handover; Accepting the completed asset; Review / Sign off and agree resources for 'Three Year In' report and recommendations
Corporate Management activities / outputs	**Vision**: Vision and Value Proposition Statement; Funding Strategy and Cash flow Optioneering Statement; Risk Probability assessment / Transfer and Assurance Model. **Risk Management**: Client Capability Statement and De-Risk Strategy; Communication and Stakeholder Alignment Strategy	**Operational Planning**: Client Side Operational Management Plan / Systems and Control Framework; Client's Requirements; Enabling, Soft and Hard Delivery; Aftercare and Operating Policy; Micro - Cash flow Planning; Procurement Strategy; Market Testing and Realignment Analysis; Manage Client Direct Responsibilities; Establish Benchmarks / Client Red Lines	**Asset Delivery**: Update Board on Design Negotiation, Impact on Business Case 'red lines' and Budget Control (Monitor Troubleshoot and Flak Activities); Review progress against Design Team Benchmarks; Updates on Design and Budget cost for Sign off; Operational / Asset Management Strategy; Design Implications Report / Change Management Implications / Risk and Opportunity Assessment; Agree Operating Policy and Plan; Agree Client Direct Works and update Design Interface Matrix to include Enabling, Fit Out and Infrastructure and other Client Interfaces	**Aftercare and Management**: Asset Handover Management Plan; Three Year In Contract Review; Three Year In Business Review
Supporting Activities	Business Process Mapping; Cognitive Mapping; Monte Carlo Analysis	Configuration and timeline; Business Delivery Planning; Update Surplus / Subsidy Analysis against actual market return		
RIBA Stage	Outline Plan Stage of Work 0-4	Outline Plan Stage of Work 0-4	Outline Plan Stage of Work 5	Outline Plan Stage of Work 6; Outline Plan Stage of Work 7
BS 6079	CONCEPT	FEASIBILITY	IMPLEMENTATION	OPERATION AND TERMINATION

Appendix 2B: RIBA Outline Plan Stage of Work

RIBA Plan of Work 2013

The RIBA Plan of Work 2013 organises the process of briefing, designing, constructing, maintaining, operating and using building projects into a number of key stages. The content of stages may vary or overlap to suit specific project requirements. The RIBA Plan of Work 2013 should be used solely as guidance for the preparation of detailed professional services contracts and building contracts.

www.ribaplanofwork.com

Stages	0 Strategic Definition	1 Preparation and Brief	2 Concept Design	3 Developed Design	4 Technical Design	5 Construction	6 Handover and Close Out	7 In Use
Core Objectives	Identify client's Business Case and Strategic Brief and other core project requirements.	Develop Project Objectives, including Quality Objectives and Project Outcomes, Sustainability Aspirations, Project Budget, other parameters or constraints and develop Initial Project Brief. Undertake Feasibility Studies and review of Site Information.	Prepare Concept Design, including outline proposals for structural design, building services systems, outline specifications and preliminary Cost Information along with relevant Project Strategies in accordance with Design Programme. Agree alterations to brief and issue Final Project Brief.	Prepare Developed Design, including coordinated and updated proposals for structural design, building services systems, outline specifications, Cost Information and Project Strategies in accordance with Design Programme.	Prepare Technical Design in accordance with Design Responsibility Matrix and Project Strategies to include all architectural, structural and building services information, specialist subcontractor design and specifications, in accordance with Design Programme.	Offsite manufacturing and onsite Construction in accordance with Construction Programme and resolution of Design Queries from site as they arise.	Handover of building and conclusion of Building Contract.	Undertake In Use services in accordance with Schedule of Services.
Procurement *Variable task bar*	Initial considerations for assembling the project team.	Prepare Project Roles Table and Contractual Tree and continue assembling the project team.	←----------- The procurement strategy does not fundamentally alter the progression of the design or the level of detail prepared at a given stage. However, Information Exchanges will vary depending on the selected procurement route and Building Contract. A bespoke RIBA Plan of Work 2013 will set out the specific tendering and procurement activities that will occur at each stage in relation to the chosen procurement route. -----------→			Administration of Building Contract, including regular site inspections and review of progress.	Conclude administration of Building Contract.	
Programme *Variable task bar*	Establish Project Programme.	Review Project Programme.	Review Project Programme.	←----------- The procurement route may dictate the Project Programme and may result in certain stages overlapping or being undertaken concurrently. A bespoke RIBA Plan of Work 2013 will clarify the stage dates and detailed programme durations. -----------→				
(Town) Planning *Variable task bar*	Pre-application discussions.	Pre-application discussions.		←-- Planning applications are typically made using the Stage 3 output. A bespoke RIBA Plan of Work 2013 will identify when the planning application is to be made. --→				
Suggested Key Support Tasks	Review Feedback from previous projects.	Prepare Handover Strategy and Risk Assessments. Agree Schedule of Services, Design Responsibility Matrix and Information Exchanges and prepare Project Execution Plan including Technology and Communication Strategies and consideration of Common Standards to be used.	Prepare Sustainability Strategy, Maintenance and Operational Strategy and review Handover Strategy and Risk Assessments. Undertake third party consultations as required and any Research and Development aspects. Review and update Project Execution Plan. Consider Construction Strategy, including offsite fabrication, and develop Health and Safety Strategy.	Review and update Sustainability, Maintenance and Operational and Handover Strategies and Risk Assessments. Undertake third party consultations as required and conclude Research and Development aspects. Review and update Project Execution Plan, including Change Control Procedures. Review and update Construction and Health and Safety Strategies.	Review and update Sustainability, Maintenance and Operational and Handover Strategies and Risk Assessments. Prepare and submit Building Regulations submission and any other third party submissions requiring consent. Review and update Project Execution Plan. Review Construction Strategy, including sequencing, and update Health and Safety Strategy.	Review and update Sustainability Strategy and implement Handover Strategy, including agreement of information required for commissioning, training, handover, asset management, future monitoring and maintenance and ongoing compilation of 'As-constructed' Information. Update Construction and Health and Safety Strategies.	Carry out activities listed in Handover Strategy including Feedback for use during the future life of the building or on future projects. Updating of Project Information as required.	Conclude activities listed in Handover Strategy including Post-occupancy Evaluation, review of Project Performance, Project Outcomes and Research and Development aspects. Updating of Project Information, as required, in response to ongoing client Feedback until the end of the building's life.
Sustainability Checkpoints	Sustainability Checkpoint – 0	Sustainability Checkpoint – 1	Sustainability Checkpoint – 2	Sustainability Checkpoint – 3	Sustainability Checkpoint – 4	Sustainability Checkpoint – 5	Sustainability Checkpoint – 6	Sustainability Checkpoint – 7
Information Exchanges *(at stage completion)*	Strategic Brief.	Initial Project Brief.	Concept Design including outline structural and building services design, associated Project Strategies, preliminary Cost Information and Final Project Brief.	Developed Design, including the coordinated architectural, structural and building services design and updated Cost Information.	Completed Technical Design of the project.	'As-constructed' Information.	Updated 'As-constructed' Information.	'As-constructed' Information updated in response to ongoing client Feedback and maintenance or operational developments.
UK Government Information Exchanges	Not required.	Required.	Required.	Required.	Not required.	Not required.	Required.	As required.

*Variable task bar – In creating a bespoke project or practice specific RIBA Plan of Work 2013 via www.ribaplanofwork.com a specific bar is selected from a number of options.

© RIBA

Index

Note: Page numbers in *italics* refer to figures and those in **bold** refer to tables.

Accelerating Change (Egan) 28
Adams, K 77
Aje, IO 62
Alderfer, CP 122
appetite for risk 218, 222
appointment of contractors *see* selection and appointment of contractors

Baban, H 89, 98–102
Baiden, BK 125
Bakar, A 120, 123, 125, 127
Banwell, H 54
behaviours of clients/construction teams 212–215
Bellemare, C 125
benchmarking 56–62, 60
bespoke nature of construction industry 21–22, 22, 150; motivation of project staff 118
bid rigging 109–110
Bigley, GA 76
blacklisting 109
Boes, H 75
Boyd, D 15, 210
Bradshaw, J 59–60
Bredillet, NB 122, 126
briefing process: checklist **14**; end users, inclusion of 13; individuals to include 14–15; information included 12–14; as ongoing 12, 20; stages 12; stakeholders, working with 13; '*voice of the client*' 12
Brooks, I 120
BS11000 31
BS6079 project cycle 212, *213*, 228
Buchanan, DA 39
Building Information Modelling (BIM) 30, 56, 67

business cases 49, 155–167, 224, 238
Business Concept Development Stage: board level ownership of risk management 246; cashflow optioneering 242–243; change and business boundaries 238; client capability statement 243–245; configuration planning 241; construction change 237, 238; corporate construction risk 243–244; de-risk strategy 243–245; funding strategy 241–242; objectives of project 240–241; organisations and planning 236–237; outputs 238–46; planned change 237; purpose 236, 246; stakeholders 244; visioning and value proposition statement 240–241
business needs, decisions based on 10–11

Carson, ER 15
Ceric, A 78, 79
Challender, J 74, 76, 78, 80–82, 117
Chan, K-Y 38, 41
Chang, S 59–60
change: in client organisations due to projects 15, 15–17, 207; construction projects as change projects 210–211; dynamic change, managing risk through 222; leadership skills of clients 40
charters in collaborative working 84, 168
checklists: compliance measures checklist 136, **137–138**; fire safety project completion checklist **145**; handover checklist **144**; master checklist 134, **135**; test certification checklist **144**
Chinyio, E 15, 210
Choi, S 61

client capability statement 243–245
client organisations: complexity of 10–12, *11*, 19–20, 149; diversity of 18; influences on 17–18, *18*, *19*; knowledge of 17–18, *18*, *19*; transformational change due to projects *15*, 15–17, 207
client requirements document 227
client-side construction risk management model *see* Construction Enterprise Risk Management (CERM) model
clients: contributions to projects 222–229, *223*, *227*, *228*; defining 8; interests represented by 8, 9; *see also* professional construction clients
codes of conduct 139
codes of ethics 110–111
collaborative working: barriers to 75; benefits of, reports of 31, 32, 150; BS11000 31; Challender research study 74, 76, 78, 80–82; charters 83, 168; context for 26, *27*; defined 74–5; expectations, agreement on 82, **83–84**, 84; government reports and initiatives 28–30, *29*; industry context 30–31; motivation of project staff 125; pre-qualification practices 67; professional context 30–31; selection and appointment of contractors 52–53, 67, 68–69; skills and qualities of construction clients 101; trust and 75–79, *77*, *78*, 80–82, 84–85
Colquitt, JA 79
communication: importance of in leadership skills of clients 39–40; as motivational factor 120–122; of requirements 8–9, 19
competitions 257
complexity of client organisations 10–12, *11*, 149
compliance measures checklist 136, **137–138**
'Constructing the Team' (Latham) 28
Construction 2025 – Industry Strategy: Government and Industry in Partnership (HM Government) 28–29
construction client model *15*, 15–16
Construction Enterprise Risk Management (CERM) model 256, 285; aftercare 253, 258, 274, 275; architecture/design team 266–267; asset delivery outputs 261–265; asset handover management planning 274–277; BS6079 project cycle comparison *228*; Business Concept Development Stage 236–246; business objectives 267; cashflow optioneering 242–243; cashflow planning and controls 253–256, *254*; change in client organisations due to projects 207; client capability statement 243–245; configuration and delivery strategy 248–249; configuration planning 241; contractor appointment 270; control of design process 250–251; Corporate Client/Delivery Capability/Transformation Stage 247–258, *254*; corporate construction risk 243–244; corporate vision outputs 240–246; de-risk strategy 243–245; defects 278; defects liability period (DLP) 275; delivery stages 227; design competitions 257; design risks and controls 267–271; enabling work 252; fit-out projects 276; funding strategy 241–242; key points 281–284; latent defects 275, 278; levels 233; milestones and timeline planning 251–252; move management 276; objectives of project 240–241; operational planning outputs 247–258; Outcomes Delivery Stage 259–271; outputs 233, 238–246, 247–258; patent defects 275; post-evaluation 278–279; post-operational review 278–279; practical completion 274–275; procurement strategy 256–257; project manager appointment 265–266; project stages, risk and 231–232; realignment analysis 257; reduction of risk, client contributions to 232–233; RIBA Outline Plan Stage of Work comparison *228*; risk management using 235; stages 233, *234*; stakeholders 243; Three Year In Stage 272–280; turnkey projects 276; visioning and value proposition statement 240–241
construction industry: bespoke nature of 21–22, *22*, 31–32, 150; reforms 23
Construction Industry Council Strategic Forum for Construction 28
Constructionline 58–59
contractor competency questionnaires 169–174
Contractor Performance Assessment Reporting System (CPARS) 59–60
Corporate Client/Delivery Capability/Transformation Stage: aftercare 253, 258; cashflow planning and

controls 253–256, *254*; configuration and delivery strategy 248–249; control of design process 250–251; design competitions 257; enabling work 252; milestones and timeline planning 251–252; operational planning outputs 247–258; outputs 247–258; procurement strategy 256–257; purpose 247; realignment analysis 257
corporate construction risk 215–219, 231, 243–244
Crespin-Mazet, F 31
Critchlow, J 52
critical success factors for construction clients: Baban research study 89, **94–97**, 98–102, **99**; collaborative working 101; conflict, dealing with 90; environment and external factors 101; human/technical balance 90, 91, 98–102, **99**, 151; importance of 89; key factors 98–102; key human skills 102–103; people skills 102–103; personal qualities 103–104; planning and organisational skills 103; project management comparison 91–92; Three Skill Approach theory 92–93, *93*, **94–97**
culture: change due to projects *15*, 15–17, 207; design and site meetings 271; motivation of project staff 124–126, 129; professional ethics 112–113, 152

de-risk strategy 243–245
decision-making process: business cases 49, 155–167; gateway approval process 46, 48–49, *50*, 153–154; levels of authority example **48**; project controls 45; project/programme boards 45–46, *47*, 49
defects liability period (DLP) 275
delivery model 227, 229
Deming, WE 57
design: briefs, clarity and unambiguity of 8–9, 19; client responses to 212–213; competitions 257; integration with construction 26, 27; professional construction clients 225–257; separation from construction 24, 54
diversity of client organisations 18
Doree, A 75
Drasgow, F 38, 41
drivers for project initiation 211–212
Dubois, A 75
Dulewics, V 104

Dvir, D 103
Dwivedula, R 122, 126

Egan, J 28, 52, 57, 58, 125, 129
Elia, D 102–103
Emerson, RW 117
end users, engagement of during projects 13, 20, 150
enterprise risk management 215; business case 224; client requirements document 227; clients' contributions to projects 222–229, *223*, *227*, *228*; design process 225–227; objective setting 223–224; operational management 224–225, 229; project success/client satisfaction 229; value propositions 223–224; *'voice of the client'* 225–227, 230; *see also* Construction Enterprise Risk Management (CERM) model
environmental ethics 115
ethics, professional: codes of conduct 113–114, **114**; codes of ethics 110–111; cultural change in the industry 112–113, 151–152; definitions of ethics 107–108; environmental ethics 115; governance and regulation 113–114; importance for construction clients 108–110; public perception of industry 109; unethical practices 109–110, **111**, 111–112
'Expectancy' theory 122
expectations, agreement on 82, **83–84**, 84
experience, clients', variety of *11*, 11–12

Farmer, M 52, 55
Festervard, D 108
financial incentives for motivation of project staff 126–127
financial model for pre-qualification 63–64
financial processes: business cases 49, 155–167; control over 49; gateway approval process 46, 48–49, 153–154; levels of authority example **48**; project/programme boards 45–46, *47*
fire safety project completion checklist **145**
Fisher, O 38, 101, 104
fit-out projects 276
Flood, RL 15
Fong, P 61
Francisco, R-V 63
Francke, A 38
fuzzy set model for pre-qualification 63

Gadde, LE 75
Galinsky, A 39
gateway approval process 46, 48–49, 50, 153–154, 155–167
Geoghegan, L 104
Ghaffarianhoseini, A 56
Gismondi, A 62, 66
Glenigan 58
governance: business cases 49, 155–167; decision-making process 49; gateway approval process 46, 48–49, 50, 153–154; importance of 150; levels of authority example **48**; professional ethics 113–114; project controls 45; project/programme boards 45–46, 47, 49
government reports and initiatives 28–30, 29
Grashina, MN 41
Guillen, L 41
Gulf Construction 52
Gurland, ST 125
Gustavsson, T 56

Hall, RJ 36, 38
Hampden-Turner, C 91
handover checklist **144**
Hatush, Z 62, 63
Haughey, D 103
health and safety handbooks 139, 175–183
Herzberg, F 117, 126–127
Holt, GD 61, 125
Huang, Y 52
Huczynski, AA 39
Hughes, W 55

Infrastructure Cost Review (HM Treasury) 29–30
inspections of sites 139
'Intelligent Client, The' (HM Government) 74, 77
'iron triangle' 22, *22*, 91, 101

Jefferson, Thomas 107

Kamara, JM 8, 9, 214
Kaming, PF 119, 123, 125, 126
Katz, RL 92–93, 100, 104
key performance indicators (KPIs) 57–61, 60, 67–68, 69
Kilduff, G 39
Knippenberg, B v 36–37
knowledge of client organisations 17–18, *18*, *19*
KPI Working Group 58

Kumaraswamy, MM 61
Kwakye, AA 52

Lam, CF 125
Laryea, S 55
latent defects 275, 278
Latham, M 28, 52, 55, 57, 58, 62–63, 125, 129
leadership skills of clients: adaptability of styles of leadership 41; challenges for 37; change process in organisations 40; characteristics and qualities 36–37; collective identity 40; communication, importance of 39–40; components of 37–38; as critical success factor 103; definitions of leadership 38–39; development of 36, 39; emergence of leaders 40; failure of projects and 41–42; (in)experienced clients 36; leadership identity development (LID) model 35, 36, 150; power, use of 40; styles of leadership 38–39, 41, 92; task/people focus 39
Lere, JC 108
lessons learnt proforma 142, **146–148**
Lester, A 38, 92
Lewis, JP 38, 91, 92, 101
Ling, FY-Y 100
Liu, AMM 108, 109, 112, 113–114
Lopes, PN 36, 37
Lord, RG 36, 38
Low Carbon Construction Final Report (HM Government) 29

management tools 211–212
Manley, K 126, 127
master checklist 134, **135**
Maurer, I 23–24
Mayer, RC 76
McCabe, S 56
Miscenko, D 38, 41
Morgan, S 31
Morote, A 63
motivation of project staff: awareness of importance of 129; benefits of 127–128, 152; bespoke nature of construction industry 118; collaborative working 125; communication 120–122; cultural change in the industry 129; cultural factors 124–126; definition of motivation 120; demotivating factors 125; equal treatment of staff 121; factors influencing **119**, 119–120, 128; financial incentives

126–127; importance of for construction clients 117–118; incentive related measures 124; job rotation to vary work 122; justification for 128–129; management style 124; playing down importance of 120–121; poor, impact of 129; pride in work 124; responsibilities assigned to staff 125; teamwork, sense of 124, 125; training and education 122–123; as undervalued 118
move management 276
Muller, R 38
multi-criteria model for pre-qualification 62–63

National British Standard (NBS) for pre-qualification 69
Nesam, L 125
Newell, MW 41
Ng, S 52, 59, 62, 63
Northouse, PG 36, 38

objective setting 223–224
Ogunsemi, D 62
Olomolaiye, PO 118, 126
operational management 224–225, 229
organisational change due to projects 15, 15–17, 207
organisational factors in success of projects 16
organisational skills 103
Outcomes Delivery Stage: architecture/design team 266–267; asset delivery outputs 261–265; business objectives 267; contractor appointment 270; design risks and controls 267–269; direct responsibilities of client 260; project manager appointment 265–266; purpose 259, 271; role of client 259–260

partnering *see* collaborative working
Patel, B 57
patent defects 275
Pearce, JL 76
performance measurement 57–58, 69
Permits to Work 136
Pinto, JK 77
planning and organisational skills 103
Portier, P 31
post-operational review 278–279
pre-qualification processes 52–53, 61–64, 66–69

pre-questionnaires 134–136, 169–174
Price, ADF 118, 126
Procurement Group 52
procurement process: alternative methods 28–31, 29; barriers and problems in traditional 23–24; car purchase comparison 21–22; collaborative working, context for 26, 27; deficiencies with traditional 24–26, 25, 32; design, separation from construction 24, 54; government reports and initiatives 28–30, 29; 'iron triangle' 22, 22; temporary basis of teams 21; *see also* collaborative working; selection and appointment of contractors
professional construction clients: business case 224; clarity of purpose 223–224, 229; client requirements document 227; contributions to projects 222–229, 223, 227, 228, 229–230; de-risking of project 221; design process 225–227; dynamic change, managing risk through 222; need for 219; objective setting 223–224; operational management 224–225; realism and flexibility 224–225, 229; role of 221; value propositions 223–224; '*voice of the client*' 225–227, 230
professional ethics: boundaries of ethical behaviour 112; codes of conduct 113–114, **114**; codes of ethics 110–111; cultural change in the industry 112–113, 151–152; definitions of ethics 107–108; environmental ethics 115; governance and regulation 113–114; importance for construction clients 108–110; public perception of industry 109; RICS codes of conduct 113, **114**; unethical practices 109–110, **111**, 111–112
programme boards 45–46, 47, 49
project boards 45–46, 47, 49
project controls 45
project development, client behaviour in 212–213, *213*
project execution plans (PEPs) 139, 184–204
project management 91–92
project manager appointment 265–266
project team, clients as integral members of 6
Publicly Available Specification 91 (PAS 91) questionnaire 59

qualities of construction clients *see* skills and qualities of construction clients

Rahman, HA 111
realignment analysis 257
reasons for project initiation 211–212
recruitment and selection of contractors *see* selection and appointment of contractors
Reeve, P 59
Reichers, AE 126
requirements, clients': communication of 8–9, 19; *'voice of the client'* 9–10
Rethinking Construction (Egan) 28
RIBA Outline Plan Stage of Work 212–213, 214, 228, 286
Rigby, J 56–57
risk: allocation of 55; Building Information Modelling (BIM) 56; identification 55; selection and appointment of contractors 55–56
risk management model for clients: allocation of risk 218; appetite, risk 218, 222; business idea as basis for projects 219; change in client organisations due to projects 207; corporate construction risk 215–219; de-risking 218, 219, 221; dynamic change, managing risk through 222; enterprise risk management 215; framework for professional clients 219; high risk for clients 207; naïve/novice, risk 218, *218*; principles for clients 215–219, *216*, *218*; professional clients, need for 219; responsibility of clients, success as 219; *see also* Construction Enterprise Risk Management (CERM) model; enterprise risk management
Ritz, GJ 100, 103
Rose, T 126, 127

Saad, M 56, 57
Schneider, B 126
'scope creep' 9
selection and appointment of contractors: benchmarking 56–62, *60*; centralised assessment systems 59–60; collaborative working 52–53, 67, 68–9; Constructionline 58–59; criteria for selection 52, 61–62; financial model for pre-qualification 63–64; fuzzy set model for pre-qualification 63; historical perspective 54–55; importance of 52–53, 150–1; key performance indicators (KPIs) 57–61, *60*, 67–68, 69; multi-criteria model for pre-qualification 62–63; National British Standard (NBS) for pre-qualification 69; performance measurement 57–58, 69; pre-qualification processes 52–53, *53*, 61–64, 66–69; problem re. collaborative working 53; Publicly Available Specification 91 (PAS 91) questionnaire 59; recommendations for improvement 69; risk 55–56; Taylor research study 51, 53, 64–69
Simon Report 54
site inspections 139
skills and qualities of construction clients: Baban research study 89, **94–97**, 98–102, **99**; collaborative working 101; conflict, dealing with 90; environment and external factors 101; human/technical balance 90, 91, 98–102, **99**, 151; importance of 89; key human skills 102–103; leadership 103; people skills 102–103; personal qualities 103–104; planning and organisational skills 103; project management comparison 91–92; success factors 93, 98–102; Three Skill Approach theory 92–93, *93*, **94–97**; training and support 91
Skitmore, M 62, 63, 110, 112, 113
staff development: construction clients 91; motivation of project staff 122–123
stakeholders: Construction Enterprise Risk Management (CERM) model 244; working with during briefing process 13
Steers, RM 120
Stewart, S 113
success of projects: enterprise risk management 229; organisational factors 16; as responsibility of clients 219; trust and 76–78, *77*, *78*; *see also* critical success factors for construction clients
Sweetland, Ben 91
systems dynamics model 15

Tabassi, A 21, 118, 120, 123, 125, 127
Taylor, S 51, 53, 64–68
test certification checklist **144**
Thompson, P 3
Thorgren, S 76
Three Skill Approach theory 92–93, *93*, **94–97**

Three Year In Stage: aftercare 274, 275; asset handover management planning 274–277; defects 275; defects liability period (DLP) 275; fit-out projects 276; latent defects 275, 278; move management 276; patent defects 275; post-evaluation 278–279; post-operational review 278–279; practical completion 274–275; purpose 272, 279; turnkey projects 276

toolkit for construction clients: after appointment of contractors 139; aims of 132–133, 149, 152; codes of conduct 139; compliance measures checklist 136, **137–138**; during construction 139–140, **141–142**, *143*; contractor competency questionnaires 169–174; cost plans **141–142**; document/process management 139–142; feedback and evaluation of 133–134; fire safety project completion checklist 140, **145**; handover checklist 140, **144**; health and safety handbooks 139, 175–183; lessons learnt proforma 142, **146–148**; master checklist 134, **135**; performance monitoring 134; Permits to Work 136; planning and devising 133; post-construction 140, 142, **144–148**; potential of 148; pre-construction documents 134–138, **137–138**; pre-questionnaires 134–136, 169–174; project dashboard template *143*; project execution plans (PEPs) 139, 184–204; reports, monthly project 139–140, **141–142**; site inspections 139; test certification checklist 140, **144**

Tourish, D 40

training and education: construction clients 91; motivation of project staff 122–123

transformational change due to projects 15, 15–17, 207

Trompenaars, F 91

trust: benefits for construction clients 79; building and retaining 23–24, 76, 80–82, 84–85; Challender research study 74, 76, 78, 80–82; charters 84; collaborative working and 75–76, 76–79, *77*, *78*; deficiencies with traditional procurement process 24–26, *25*, 151; definitions 76; expectations, agreement on 82, **83–84**, 84; success of projects and 76–78, *77*, *78*; unknown factors in construction process 22

Turner, R 38

turnkey projects 276

Turskis, Z 52, 53

value propositions 211–212, 223–224, 240–241

Vass, S 56

Vee, C 110, 112, 113

visioning and value proposition statement 240–241

Vitell, C 108

'voice of the client' 9–10, 12, 19, 149, 225–227, 230

Vroom, VH 122

Walker, A 38, 39, 92

Wei, L 62

Wilkinson, I 52

Wolstenholme, A 55

Wong, Z 100, 103

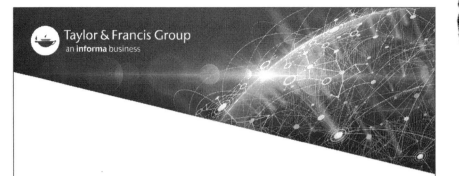